Stereotaxic Neurosurgery in Laboratory Rodent

Barbara Ferry • Damien Gervasoni
Catherine Vogt

Stereotaxic Neurosurgery in Laboratory Rodent

Handbook on Best Practices

Springer

Barbara Ferry
Lyon Neuroscience Research Centre
CNRS UMR 5292–INSERM U 1028
Université Claude Bernard Lyon 1
Lyon
France

Catherine Vogt
Université Claude Bernard Lyon 1
Lyon
France

Damien Gervasoni
Lyon Neuroscience Research Centre
CNRS UMR 5292–INSERM U 1028
Université Claude Bernard Lyon 1
Lyon
France

Translation by Dr. Marc Marien from the French language edition 'Précis de neurochirurgie stéréotaxique appliquée aux rongeurs de laboratoire' © Editions Lavoisier, Cachan, 2012; ISBN : 9782743014599

ISBN 978-2-8178-0471-2 ISBN 978-2-8178-0472-9 (eBook)
DOI 10.1007/978-2-8178-0472-9
Springer Paris Heidelberg New York Dordrecht London

Library of Congress Control Number: 2014931789

Printed on acid-free paper

Springer is part of Springer Science+Business Media (www.springer.com)

Foreword

The use of animals for research purposes can only be justified under a set of strict conditions. During the last decade, animal experimentation has become more and more subjected to the control of legislative regulations.

The European Directive 2010/63/EU on the protection of animals used for scientific purposes requires that experiments be performed solely by competent, authorized persons who are adequately trained (Art 23). According to this Directive, nonbinding guidelines at the level of the Union on the requirements laid down in Art 23 – at the time of writing submitted – may be adopted in accordance with the advisory procedures referred to in Art 56(2). Research institutions nevertheless must provide training on humane care and use of animals, on methods to minimize pain, suffering, distress, or lasting harm, on design of procedure and projects, and on implementation of the 3Rs (Reduction, Refinement, Replacement). Most of these "core" modules do not deal with surgical techniques which is understandable, since only a limited number of persons are involved in research projects which include surgical procedures.

For this latter group, however, there is an evident need for more advanced training. For the reader interested in stereotactic surgery in the rodent, this handbook may be seen as an important teaching tool.

The first recorded instance of the use of rats in scientific research was in 1856 in a paper by J.M. Philippeaux entitled "note sur l'extirpation des capsules surrénales chez rats albinos" (C.R. Acad. Sci, Paris, 43, 904). Since then, rodents have undoubtedly remained the premier warm-blooded vertebrate in experimental work (EU statistics 2008).

There is no mystery as to why rodents command such a high standing in experimental work. They breed easily, profusely, and continuously during their normal reproductive lifespan in the laboratory and, because of their small size, can be housed in large numbers in a relatively small area and, with practice, can be easily handled. Much is known about their physiology, anatomy, genetics, and behavior, and meaningful results can be obtained in rats which, if interpreted with care, can be extrapolated to man.

Stereotactic surgery began in 1908, when Sir Victor Horsley and Robert Clarke described and introduced their new apparatus which allowed them to introduce a probe, blade, or needle under accurate control into a subcortical structure of a monkey. Horsley and Clarke indeed described the first animal stereotactic apparatus, by which Cartesian coordinates (x, y, and z) could be assigned to every point within the brain. Their system was based on three Cartesian planes (midsagittal; basal, running though the external auditory canals; and coronal, perpendicular to the two other planes and also running through the external auditory canals). Such stereotactic systems allowed investigators to insert an electrode or cannula into a desired subcortical target area, with minimal damage to the overlying cortex and white matter.

This handbook is intended as an introduction to stereotactic surgery in rodents for the new researcher, and it is written in a way which will, it is hoped, allow it to be used as a laboratory manual. To this end, it is organized into seven main sections with the first few providing important background to the rest of the book.

No introduction into the use of animals for experimental purposes would be complete without encouraging the reader to use a humane approach in the use of animals. In practical terms, it means treating animals compassionately, having regard to their welfare, and, in particular, reducing all unnecessary pain and suffering. An update of actually applicable legislation and ethical considerations are clearly laid out in Chapter One.

The following chapters place particular attention on descriptive and functional neuroanatomy, description of elements to take into account to ensure the perioperative care and surgery, justification for experimental stereotactic surgery, and its follow-up.

The authors hopefully will succeed in convincing the reader of the important rule that a surgically prepared animal should not be allowed to suffer postoperatively from stress derived from an incompetent technique or an ignorance of surgical care, all too possible if the investigator is ill-concerned with the fundamentals on which proper surgery is based.

Pain and discomfort are not always avoidable in experiments that involve surgical procedures. Although this book does not deal with the moral aspects of animal experimentation, an ethical judgment prior to the performance of the experiments is considered by the authors as a matter of extreme importance. Also it is seen *as a condition sine qua non* that persons involved in experiments are adequately trained. I can only say that I am fully in line with the authors when stating that this book aims to improve competency in order to contribute to a more responsible use of animals in neurobiological research.

As coauthor of the 1991 revised "Red Book" "Guidelines for the Care and Use of Mammals in Neuroscience and Behavioral Research," NRC, 2001, providing information on best practice for all mammalian species in neuroscience research, I welcome this handbook as an attempt at a remedy, trying to enlighten the young and even more experienced scientist with some knowledge of stereotactic techniques in rodents; it should trigger the reader to look for more specific guidance in the topic he or she wants to.

Castres, France Anne-Dominique Degryse

CV of Anne-Dominique Degryse

Dr. Anne-Dominique Degryse is Head of Laboratory Animal Resources of CRPF, Institut de Recherche Pierre Fabre (IRPF), Castres, France, since 1991. She obtained her veterinary degree from the Veterinary School in Ghent, Belgium, in 1980. From 1980 to 1986 she worked at Janssen Pharmaceuticals in Beerse, Belgium, as a Researcher in the Veterinary Department. From 1987 to 1991 she was Head of the Animal and Pretest Unit of the Toxicology Center of Merck Sharp & Dohme-Chibret in Riom, France. She joined Pierre Fabre in 1991. Dr. Degryse has coauthored more than 80 publications relative to laboratory science and medicine in specialized international peer-reviewed journals or chapters in books. On top of her daily activities as Head of Laboratory Animal Resources, CRPF, and as Director of Coordination of Laboratory Animal Science and Welfare of the IRPF, she fulfills other responsibilities to include the following.

In 1997, she cofounded the European Society of Laboratory Animal Veterinarians (ESLAV), of which she currently serves as the past President. In addition, she is a Diplomate of the European College of Laboratory Animal Medicine (ECLAM) of which she currently serves as the President and a member of the Training Committee, which approves ECLAM training programs. Also, she represents ECLAM as a member of the European Board of Veterinary Specialisation (EBVS), and in 2004, she was awarded the title of Honorary Diplomate of the American College of Laboratory Animal Medicine (ACLAM).

In 1997, she was the first European to be nominated on the Council on Accreditation (COA) of the Association for Assessment and Accreditation of Laboratory Animal Care International (AAALAC International). She has served AAALAC International ever since. She was appointed Section Leader to the European Section of the COA, a position she held from 2004 to 2007, and COA member until 2009.

From 2009 to 2011, she served as Council Emeritus Member and in 2011 was elected as a Member at Large of the Board of Trustees (BOT) and Member of the Executive Committee. Dr. Degryse is a member of numerous French and European governmental and nongovernmental organizations active in the field of laboratory animals. In 2010 she was nominated as Convener for the FELASA Working Group on Severity Classification but for personal reasons was replaced by Dr. David Anderson in this role.

In 2011, at its annual meeting of the BOT, AAALAC International presented its highest honor, the Bennett J. Cohen Award, to Dr. Degryse. The Bennett J. Cohen Award recognizes outstanding individuals who have provided exceptional service and significant contributions to AAALAC International and have demonstrated a strong and abiding commitment to advancing science through the promotion of the highest standards of laboratory animal care in research, testing, and education.

Dr. Degryse was awarded the very prestigious Charles River Prize 2013 during the National AALAS Meeting in Baltimore, Maryland, USA, in october 2013.

Preface

The neurosciences have evolved considerably over the course of recent decades, from a conceptual point of view inasmuch as a methodological one. Our knowledge of the central nervous system and its functions has increased in a significant manner, thanks to modern functional imaging techniques.

With the development of microdialysis and multielectrode electrophysiology, and the more recent methods of optogenetics, invasive techniques for the exploration of the brain remain current and more than ever indispensable for perfecting our knowledge of the architecture of the central nervous system and its pathways.

Stereotaxic surgery is rarely if ever taught as a subject per se within an educational curriculum. Rather, this knowledge and the accessory know-how and skills are most often acquired and passed on within a working research laboratory setting. Depending on the tools and equipment available, this modus operandi results over time in a wide variety of practices, certainly effective since they are often invented and developed in order to respond to original and specific research problems, but often hindered by the constant evolution in knowledge and regulations related to animal experimentation.

In this context, the ambition of this book was to consolidate the essential concepts of neuroanatomy, position (or coordinate) space, anesthesiology, and actual surgery to better guide the research investigator in their options and to assist him or her in the ethical practice of stereotaxic surgery on laboratory rodents. Even with textbooks and original scientific publications serving as indispensable sources of information in any laboratory's library, it seemed important to us to create a "workbench" manual for investigators, simpler and better suited to the preparation and execution of stereotaxic procedures.

The difficulties in this undertaking were to remain concise while probing deeply into certain topics, which obliged us to make certain choices in the themes covered and to organize the material in a readable manner, like a step-by-step "methods" manual.

Fruit of a collective effort, the choices reflect the background, basic training, and specialty of the authors and their experience from daily practice. Beyond being a simple practical guide, we hope that this book will provide food for thought and will encourage students, technicians, engineers, and scientists in their research.

After an introduction focusing on regulatory and ethical considerations, a review on descriptive and functional neuroanatomy (including a non-exhaustive list of the major neurotransmitter systems in rodents and the experimental tools used to modify their activity), the book addresses the more practical aspects of stereotaxic neurosurgery.

From a description of stereotaxic atlases to the handling and operation of existing stereotaxic instruments, the investigator is guided through the utilization of stereotaxy to define coordinates for target structures in the central nervous system.

Much of the book is then dedicated to the technical gestures that have to be carried out before and during the procedure so that the target structure is attained with a maximum of precision. An entire chapter notably addresses the preoperative phase with all the procedures that precede the actual surgery: the verification of the health status of the animals, the preparation and sterilization of surgical equipment (insisting here on the respect for the principles of asepsis), the different ways for reducing pain during and after the surgical procedure, as well as the implementation of an appropriate anesthetic protocol. This is followed by a list of suture types that can be performed at the end of the procedure.

The subsequent chapter covers the practical realization of a stereotaxic surgery through three examples of procedures, with many references to recent publications dealing with modern methods of the stereotaxic approach.

Finally, the last chapter provides detailed descriptions of autopsy and histology protocols.

Covering some common applications of stereotaxy in the rat and mouse, the book also contains appendices allowing even the novice investigator to carry out an entire stereotaxic surgical procedure, respectful of current ethical policies and in accordance with the legislation in effect.

Today the technique of stereotaxic neurosurgery is used widely in public and private research laboratories, by a diverse population of scientists, engineers, technicians, students, and animal care staff.

To meet the new requirements in matters of ethics in animal experimentation, and in particular to implement the rule of three “Rs” (Replacement, Reduction, and Refinement), it is essential to precisely define a working framework that outlines the procedures and conduct to follow during the course of an experiment involving stereotaxic surgery.

Thus, it is essential for those in the scientific community who practice stereotaxic surgery, now or in the future, to be able to refer to a unique document that integrates an ensemble of standardized procedures compliant with current European directives regarding laboratory animal welfare. This manual is such, a vade mecum, a handbook providing a detailed review of the gestures that must be rigorously executed during and after the surgical phase of an experiment.

Lyon, France Barbara Ferry
Lyon, France Damien Gervasoni
Lyon, France Catherine Vogt

Acknowledgments

We hereby would like to express our warmest appreciations and thanks to Dr. Marc Marien who has graciously consented to translate this book into English.

None of our objectives would have been reached without the skill, talent, and dedication of Marc, and we are grateful for his high professionalism, availability, and for the kind willingness he always demonstrated throughout this laborious task.

We also want to express our profound thanks to Anne-Dominique Degryse who has honored us with the time she took to read and prepare the foreword to this book.

Barbara Ferry
Damien Gervasoni
Catherine Vogt

Contents

Introduction

The word "stereotaxy" is derived from the Greek *stereos*, meaning solid, three-dimensional, and *tactos*, meaning ordered or arranged, and thus denotes the concept of "arrangement in space."

Stereotaxic Neurosurgery

With the development of histological techniques for central nervous system tissue, neuroanatomists were able to begin describing the complex anatomical connections within and between cerebral structures. In 1887, Sir Victor Horsley and his collaborator R.H. Clarke, in their studies of the connections of the cerebellum in the rhesus monkey, realized that it was necessary to develop a method that could allow the precise lesioning of the different cerebellar nuclei. This need became the mother of an invention: stereotaxy.

The brain was modelled in three planes: horizontal, sagittal, and frontal, forming a system of three-dimensional coordinates. Clarke invented an instrument that allowed the stable maintenance of the animal's head within this system of coordinates. With this new "stereotaxic device" and a microtome to cut through the cranial bone, they were able to elaborate models showing the positions of cerebral structures with respect to others, as well as their positions in relationship to the surface of the skull. With the aid of such models and the stereotaxic device, Horsley and Clarke (1908) could measure the skull of the animal and pierce an opening just large enough to permit the introduction of a lesioning electrode. In addition to allowing the precise placement of the electrode, this method afforded minimal damage to underlying cerebral regions as well as a minimal loss of bony tissue.

The principles of stereotaxic surgery used today are essentially the same as those pioneered by Horsley and Clarke. Although the stereotaxic apparatus was first developed at the beginning of the twentieth century, it was not until 1947 that a similar type of instrument was adapted for human neurosurgery. Before the era of magnetic resonance imaging (MRI), stereotaxy required conventional radiological equipment for visualizing the cranial vault and its contents and arteriographs and ventriculographs to control for the positioning of instruments with respect to the target zone and to reduce the risks of hemorrhage associated with imprecise surgical manipulation. The appearance of the scanner and later the technique of MRI significantly modified stereotaxic procedures, particularly in regard to tumoral pathologies (Ferry et al. 1988; Sedan et al. 1987) and to the materials used in the fabrication of stereotaxic frames according to the constraints of this type of imaging. Modern numeric-based imaging is at the foundation of the computerization of stereotaxic approaches, the storage of digitalized images, calculations of target coordinates, laser-guided stereotaxy, control of the operating microscope, and the robotization of surgical procedures (Kelly 1991).

Human stereotaxic surgery has well established itself over the last two decades. Its principle applications today include the symptomatic treatment of motor disorders (in particular parkinsonian syndromes, thanks to functional stereotaxy with neuronal grafts and deep brain stimulation), biopsies of tumoral pathologies and interstitial radiotherapies, as well as the treatment of hydrocephaly, epilepsy, and cerebral arteriovenous malformations.

The integration of stereotaxic methodologies as an additional surgical gesture (neurosurgery "on frame") and the possibilities of computerization of the technique assure it a growing place in neurosurgery.

Chapter 1
Regulatory and Ethical Considerations

Abbreviations

ASPA	Animal Scientific Procedures Act
AWA	Animal Welfare Act
AWB	Animal Welfare Body
AWRs	Animal Welfare Act and Regulations
CBC	Complete Blood Count
CCAC	Canadian Council on Animal Care
EKG	Electrocardiogram
FELASA	Federation for Laboratory Animal Science Associations
g	gram
IACUC	Institutional Animal Care and Use Committee
ILAR	Institute for Laboratory Animal Research
NIH	National Institute of Health
NTS	Nontechnical Summaries
OACU	Office of Animal Care and Use
PHS	Public Health Service
UK	United Kingdom
US	United States
USDA	United States Department of Agriculture

1.1 Purpose of Law and Regulations

The use of animals in research still arouses intense debate. Its necessity, justification, and acceptability are matters of widely varying opinions, mainly based on moral convictions. There is, however, a consensus to allow animal research as long as no valid alternative method exists to reach the objective. Most countries have their own legislative framework concerning animal research, and this chapter briefly presents examples of laws and regulations, some of which are currently being

B. Ferry et al., *Stereotaxic Neurosurgery in Laboratory Rodent*,
DOI 10.1007/978-2-8178-0472-9_1,

updated. The purpose of this non-exhaustive enumeration is not to compare their respective strengths or to highlight a country having prior claim in this matter but to illustrate concepts that are common to most legislation and to provide the reader with references and resources.

All researchers, regardless of their area of specialty, should be aware of the laws and policies on the use of animals for experimental and other scientific purposes in force in their country. Besides the legal aspect of animal experimentation, users are strongly advised to follow the latest principles and guidelines for animal care and use available in their own institution, in particular on anesthesia and humane euthanasia procedures. Interactions with local animal welfare bodies such as Institutional Animal Care and Use Committee (IACUC), a referent veterinarian, together with a proactive search of the latest available knowledge or alternative method should be encouraged. Valuable publications and resources are available and regularly updated. These include the guidelines and recommendations from the Federation for Laboratory Animal Science Associations (FELASA) (Voipio et al. 2008; Guillen 2012), the guidelines from the Institute for Laboratory Animal Research (ILAR) (Committee on Recognition and Alleviation of Pain in Laboratory Animals et al. 2009; Committee for the Update of the Guide for the Care and Use of Laboratory Animals et al. 2011),[1] the Canadian research Council on Animal Care (CCAC) in Science (2013a, b),[2] and the website of the UK National Centre for the Replacement, Refinement and Reduction of Animals in Research (NC3Rs).[3]

With the revised Directive 2010/63/EU for the protection of laboratory animals[4] that took effect on January 1, 2013, Europe has recently reinforced the implementation of the principle of the three Rs (Replacement, Reduction, and Refinement), originally put forward by Russell and Burch in their "The Principles of Humane Experimental Technique" (Russell and Burch 1959)[5] on the use of animals for experimental and other scientific purposes (see also Appendix 6). While its ultimate goal is to replace the use of animals, this new Directive acknowledges that animals are still needed for research, especially basic research. As in the preceding Directive (86/609/EEC), an appropriate education and training is required for all those engaged in the use of live vertebrate animals for scientific purposes. The new Directive 2010/63/EU distinguishes the following functions: (a) carrying out procedures on animals, (b) designing procedures and projects, (c) taking care of animals,

---

[1] Both references are freely available from the website of the National Academies Press at http://www.nap.edu.

[2] Note that Volumes 1 and 2 of the Guide to the Care and Use of Experimental Animals are out of date and currently under revision. Chapters XIX (laboratory mice) and XXI (laboratory rats) of Volume 2 are available in a PDF format with some sections replaced by references to more recent CCAC publications where available.

[3] http://www.nc3rs.org.uk/.

[4] The Directive 2010/63 is available in several languages on the EU's portal on law website at eur-lex.europa.eu/ or on the European Commission website http://ec.europa.eu/environment/chemicals/lab_animals/legislation_en.htm.

[5] A PDF version of the second edition (1992) is available online at http://altweb.jhsph.edu/pubs/books/humane_exp/het-toc, accessed September 10, 2013.

and (d) killing animals (Article 23, Directive 2010/63/EU).[6] Persons involved in tasks a, c, and d must be supervised until they have demonstrated the requisite competence; persons in charge of designing procedures and projects (b) must have received instruction in the relevant scientific discipline and have species-specific knowledge. Individuals working with animals should thus have and maintain state-of-the-art knowledge and skills concerning the constant improvements in laboratory animal sciences and animal use in research. The minimum requirements with regard to education and training are enumerated in Annex V of the Directive 2010/63/EU. Appropriate qualification and experience are especially required for surgery. The new legislation updates the minimum standards for animal housing and training of users and aims to improve animal welfare. The revised Directive notably introduces measures that strengthen the evaluation of the need of animal use: if not already common practice, getting approval by an Ethics Committee on animal use prior to the work commencing is now mandatory in the 27 member states. According to the terms of the revised Directive (Article 36–41, Directive 2010/63/EU), authorization for the conduct of a project involving live animals in experimental procedures shall not be granted without a positive ethical review.

The ethical review process requires from experimenters a prospective assessment of the severity of the procedure (Article 15, Directive 2010/63/EU), which may be either "non-recovery," "mild," "moderate," or "severe." In addition, Article 54 on reporting requires that for the purpose of statistical information, the actual severity of the pain, suffering, distress, or lasting harm experienced by the animal must be reported. The actual severity of any procedure may be different from the prospective assessment or the prediction of severity made at the time of the project evaluation, but in all cases, it will be a key consideration in determining whether an animal can be reused in further procedures (Article 16, Directive 2010/63/EU).

The new European legislation has also implemented changes in the organization of the institutions in charge of animal welfare in the member states. In each establishment, a local animal welfare body (AWB) is created that shall include at least one person responsible for the welfare and care of the animals and an identified user, a scientist. The AWB is principally in charge of following the development and outcome of research projects, establishing and reviewing operational procedures such as monitoring or reporting, and advising the staff dealing with animals on matters related to the welfare of animals, their acquisition, accommodation, and care. Importantly, the AWB shall be in touch with a designated veterinarian with expertise in laboratory animal medicine (or a suitably qualified expert where more appropriate). In each member state, a competent authority is designated to be in charge of delivering an authorization for the project and for carrying out regular inspections. If not already in place, governmental agencies or advisory committees are created to monitor practices of animal use in research, to gather any elements

[6] The functions a, b, c, and d identified in the Directive 2010/63/EU should not be confused with FELASA categories A, B, C, and D (Category A, persons taking care of animals; Category B, persons carrying out animal experiments; Category C, persons responsible for directing animal experiments; Category D, laboratory animal science specialists).

that may contribute to the further implementation of the requirement of replacement, reduction, and refinement. For instance, in the UK, the Animal Procedures Committee created after the Animals (Scientific Procedures) Act 1986 (ASPA)[7] and charged with advising the British Home Secretary on matters related to animal research is currently being replaced by an Animals in Science Committee.[8] In France, the Ministère de l'Enseignement Supérieur et de la Recherche is now in charge of granting the authorization of projects involving animals. All experimenters are therefore advised to identify and become acquainted with the various authorities and other entities as well as their missions and responsibilities.

In the USA, the Animal Welfare Act (AWA)[9] signed into law in 1966 and enforced by the Animal Care Agency of the US Department of Agriculture (USDA) establishes the minimum acceptable standard and sets the requirements for the conduct and control of the use of animals in research. In its latest version, the AWA is a reference for all other laws, policies, and guidelines for animal care and use (e.g., Public Health Service (PHS) Policy on Humane Care and Use of Laboratory Animals (Office of Laboratory Animal Welfare and National Institute of Health, 2002)). Useful resources derived from the AWA can be found in the Guide for the Care and Use of Laboratory Animals (Committee for the Update of the Guide for the Care and Use of Laboratory Animals et al. 2011)[10] and the Guidelines for the Care and Use of Mammals in Neuroscience and Behavioral Research (Committee on Guidelines for the Use of Animals in Neuroscience and Behavioral Research et al. 2003).

Legislative frameworks in most countries are the result of large-scale consultations of committees composed of representatives from academia, research organizations, veterinarians, as well as legal and ethical specialists and representatives of animal welfare organizations. Whether or not the research will find an outreach, all studies involving animals are now subject to a higher scrutiny prior to their beginning to ensure their necessity, proper justification, and adequate environmental and material context and design for their conduct, with an increasing attention on their compliance with the 3R principles (Fig. 1.1). For this purpose, most countries have introduced simplified administrative procedures through which investigators can obtain an agreement or license to perform animal research. Rather than a burden, this process should be seen as a step in the implementation of the 3Rs. Indeed, the submission of an animal protocol to an Animal Care Committee or equivalent authority represents a unique opportunity to thoroughly think about ways to reduce the number of animals used, to minimize animal pain and distress, and to consider alternative approaches.

---

[7] http://www.legislation.gov.uk/ukpga/1986/14/contents.

[8] Users may consult the draft of the Operation of the Animals 1 (Scientific Procedures) Act 1986 (as amended) at the link http://www.homeoffice.gov.uk/publications/science-research-statistics/animals/transposition_of_eudirective/aspa-draft-guidance.

[9] Animal Welfare Act as of February 1, 2010 (http://awic.nal.usda.gov/government-and-professional-resources/federal-laws/animal-welfare-act), accessed September 10, 2013.

[10] The eighth edition (2011) is available on the website of the National Academic Press at http://www.nap.edu/catalog.php?record_id=12910, accessed September 10, 2013.

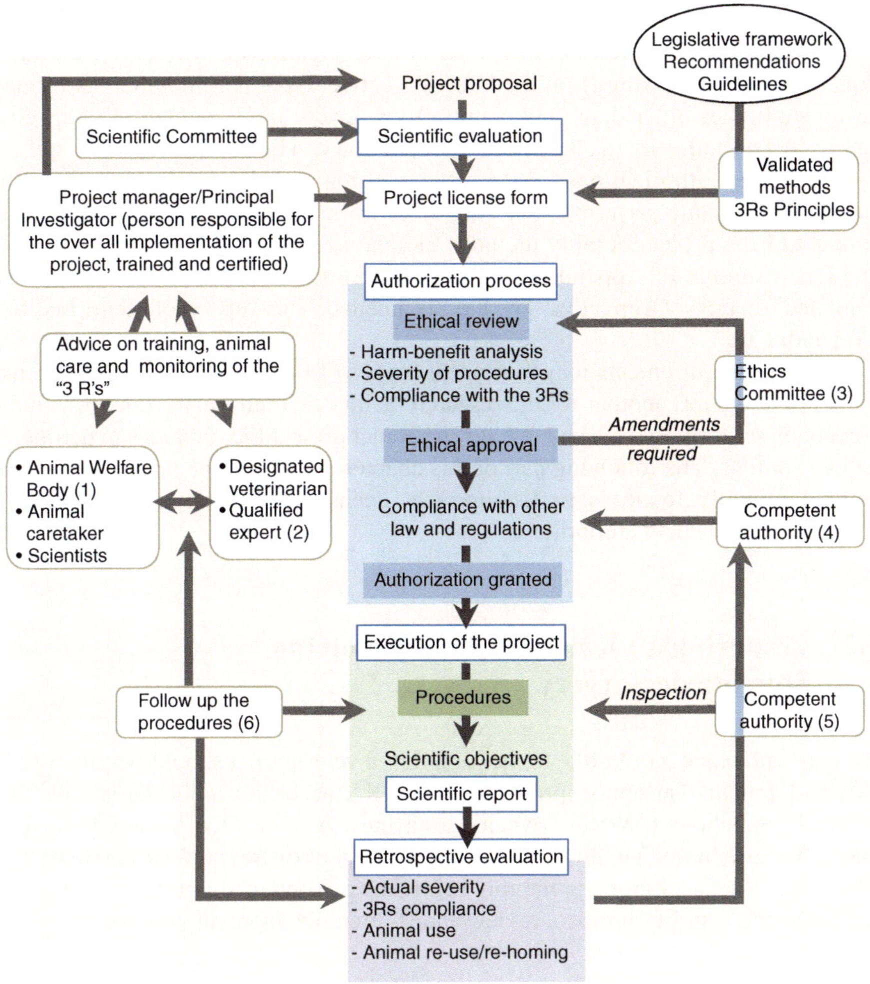

Fig. 1.1 Flowchart for the implementation of a project

Inspired by the European Directive 2010/63/UE (January 1, 2013), this flowchart may apply to countries outside the European Union with adequate transposition. The apparent complexity of the flow chart is representative of the higher scrutiny applied to research or teaching projects that use live animals. (1) According to the last Directive, a local animal welfare body (AWB) is created in each establishment to work as close as possible with scientists. The AWB is roughly equivalent to an IACUC except that it is not in charge of the ethical review, a task usually assigned to a distinct Ethics Committee. (2) The designated veterinarian or qualified expert is a person with advisory duties regarding the well-being and treatment of the animals. (3) Ethics Committees are usually composed of at least a veterinarian, an animal caretaker (zootechnician), scientists, and a nonscientist. (4) A competent authority means an authority or body designated by a member state to deliver project authorizations and to carry out additional obligations arising from the latest Directive. Note that member states may designate bodies other than public authorities for the implementation of specific tasks. (5) The authority in charge of the inspection can be distinct from the authority that grants authorizations for the project, although both entities may be affiliated with the same governmental agency or ministry. (6) The follow-up of the conduct of the project notably includes the maintenance of records for each animal (see example in Appendix 2), with a daily monitoring of the health status, pain and distress (Appendix 4), and, if needed, the report of abnormalities (Appendix 1).

Licensing requirements may imply three types of licenses, an individual license for the investigator, another for the research facility that must also comply with the guidelines set forth by law, and one for the actual project that includes experiments on live animals. The following part of this chapter deals with the latter and lists the items commonly found in protocol/project license forms that are submitted to IACUCs or equivalent authorities.

1.2 Preparing a License Protocol Involving Stereotaxic Surgery

As a preamble, we would like to mention that a veterinarian should be involved in the development of an animal protocol. In the USA, according to the Animal Welfare Act and regulations (AWRs), a veterinarian must be consulted for any procedure that may cause more than momentary or slight pain or distress (AWR 2.31(d)(1)(iv)(B)). Such veterinary input, preferably before the protocol is submitted to a committee, can often simplify protocol review and shorten the approval process.

1.2.1 Prospective Assessment of Pain and Distress

Most project license forms require a prospective assessment of the severity of the procedure (Article 15, Directive 2010/63/EU). Stereotaxic surgeries in rodents are typically procedures that may cause more than momentary or slight pain. As such, they should only be performed with appropriate sedation, anesthesia and analgesia. In general, surgical or other painful procedures are not permitted on unanesthetized animals paralyzed by chemical agents. Whether the surgery is included in a survival procedure or is a terminal one, the rank of pain/severity will be the same. In the USA, both survival and non-survival procedures refer to the same USDA[11] category

[11] US Department of Agriculture.

D: animals used for teaching, research, experiments, testing, or surgery that will involve accompanying pain or distress and for which appropriate anesthetic, analgesic, or tranquilizing drugs will be used.[12] In Europe, procedures are now classified as "non-recovery," "mild," "moderate," or "severe." The assignment of a severity degree shall be based on the most severe effects likely to be experienced by an individual animal after applying all appropriate refinement techniques. A procedure that includes a single stereotaxic surgery with a craniotomy should be ranked "moderate" (surgery under general anesthesia and appropriate analgesia, associated with postsurgical pain, suffering, or impairment of general condition (Section III of Annex VIII, Directive 2010/63/EU)). If the surgery is expected to result in severe or in moderate but persistent postoperative pain, suffering, or distress or is likely to cause severe impairment of the well-being of the animal, the procedure shall be classified as "severe" (equivalent USDA Category E[13]). For example, this can be the case if anesthetic, analgesic, or tranquilizing agents adversely affect the procedure and its results or interpretation and, therefore, cannot be used. For such procedures, the withholding of analgesia, anesthesia, or other treatment to alleviate pain/distress will have to be scientifically justified: in particular, the experimenter must provide the methods or means used to determine that pain and/or distress relief would interfere with the results. Note that for such severe/category E procedures, federal/state/local laws or policies may require separate declarations/forms and a systematic retrospective report.

Besides surgical pain, stereotaxic surgeries like other invasive techniques can induce transitory or even permanent functional damage or handicap. Depending upon the target area, or the way chosen to reach it, functional damage may be created by the sole introduction of a foreign object into the brain. To some extent, such damage can often be anticipated. For instance, if the procedure aims to lesion a brain area involved in motor control, a functional consequence should be expected that would affect locomotion or movement coordination and therefore compromise the animal's well-being. For instance, neurotoxin lesioning of dopaminergic neurons of the substantia nigra by 6-hydroxydopamine (6-OHDA) in rats (Schwarting and Huston 1996a, b; Kirik et al. 1998) induces a postural asymmetry and a unilateral akinesia when performed unilaterally. Partial or complete lesions are often accompanied by a spontaneous or induced sensitization of locomotor activity that can acutely or chronically affect the animal's ability to eat, drink, and behave normally (Ungerstedt 1971). In such cases, adequate postoperative care procedures such as manual feeding can be required until the animal recovers its autonomy and a complete restoration of its behavioral repertoire. In a new project, a preliminary study with few animals can be included to tentatively determine if the procedure can cause a potential handicap.

In general, only one major survival surgery (i.e., having the potential for producing a permanent handicap for an animal expected to recover from surgery) may be

[12] From the US Government Principles for the utilization and care of vertebrate animals used in testing, research, and training.

[13] By law, all category E procedures must annually be reported to the USDA.

performed per animal. If multiple survival procedures are essential parts of the project, experimenters are expected to provide a thorough justification and to describe the chronology of the multiple surgeries, how they are interrelated, and why they are necessary to achieve the scientific objective. Here again, cumulative pain, distress, or functional deficit that may result from the succession of procedures should also be detailed.

When assessing a level of pain and distress prior to an experiment, the entire lifetime experience of the animal should be considered, not just the animal's experience during the stereotaxic surgery. It is therefore crucial to draw attention to the other factors that can also affect the animal's well-being, such as transportation, housing, feeding and handling, and individual history if the animal has already been used in previous experiments. Procedures or treatments that precede the surgery (e.g., operant conditioning, irradiation, food/drink restriction, environmental distress, forced exercise, disease conditions, exposure to predator scents, and excessive noise) are other factors to consider. One needs to identify as many potential sources of pain and distress as possible, their intensity, and duration. In addition to the gestures performed during the surgery itself, and among other sources, incorrect tissue manipulation, inexperience of the experimenter, inadequate postoperative care, or insufficient acclimatization of the animals are potential sources of pain or distress. Furthermore, the phenotype, particularly in genetically modified animals (transgenic, knockout, knockin, floxed, cloned, or otherwise genetically modified strain) should not be neglected. For instance, some rat or mouse strains can be prone to seizures, and additional complications may arise from this condition. Listing all potential sources of pain or distress in the experimental model will not only help in the prospective assessment of pain/distress but also serve to define additional gestures and care of the subjects that will be proposed to the committee and performed by the collaborators of the study and the animal caretaker.

1.2.2 Measures to Alleviate Pain and Distress

After assessing the level of pain and distress in their models, experimenters are asked to thoroughly describe the methods and means used to monitor and to alleviate them.

1.2.2.1 Monitoring Pain and Distress

The ability to recognize the clinical signs of pain and distress in animals is fundamental to their relief. Researchers and animal care staff should therefore have a solid knowledge of the normal and abnormal appearance, behavior, and physiology of the animals. Many publications are available on health monitoring in laboratory rodents (Rehbinder et al. 1996; Nicklas et al. 2002). Below is a list of common clinical signs of pain observed in rodents (see also Appendix 3). This list is not meant to be definitive but provides some basic elements of checking the health status before a surgery and throughout the entire experimental procedure.

In its appearance, an animal in pain or distress may present an abnormal stance, an arched back, a rough or stained hair coat, or soiled orifices (diarrhea or urine). Rats can also exhibit red porphyrin staining around the eyes and nose; a rapid, shallow respiration; or tremors. Behavioral signs may include decreased general activity, lethargy or reserve to move around the cage, no curiosity, no nesting activity, absence of grooming, change in locomotion (stumbling, falling, tiptoe walking), no or limited interaction with congeners, hiding, defensive state, unusual aggressiveness, self-scratching and biting, excessive licking, or self-mutilation. Rodents are nocturnal animals, i.e., mostly active at night, so that immobility during the day may signal sleep behavior that is accompanied by a reduced reactivity to external stimuli. Decrease in food or fluid intake, a drop in body weight, or symptoms of dehydration are signs to be watchful for. A rat or a mouse previously habituated to handling may show a sudden intolerance to manipulation, a hunched posture, no righting reflex, and vocalization. A prior knowledge of each animal's normal behavior, but also of their ability to sometimes behave differently when they feel they are under close watch, will help in the observation and evaluation of their well-being. Physiological signs of well-being may include biochemical indicators that require dosage in blood samples.

In the case of survival surgery, a close postoperative follow-up is also required, which should include at least daily observations of the animal's behavior and evaluation of signs of distress. Individual logs or checklists are invaluable for this as well as a critical use of empathy. Note that keeping contemporaneous records of each animal on which procedures have been carried out will also serve to prepare a final report on animal use and to determine the actual level of pain and distress of the procedures.

1.2.2.2 Alleviating Pain and Distress

Four general approaches are available to minimize pain (Dubner 1987): the use of general anesthesia, the use of local anesthesia and/or analgesia (see Sect. 5.3), the training of animals to avoid situations that produce pain (escape-avoidance behavior), and control of the intensity and/or duration of the stimulus by the neuroscientist. In addition to the agents described in Chap. 5 and Appendices 7 and 8 of this manual, readers are strongly encouraged to get an input from their referent veterinarian and refer to the guidelines of their own institution when proposing ways and pharmacological agents to alleviate pain and distress.

1.2.3 Experimental Design and the Number of Animals

An adequate design is required for an experimental study to be valid. How many animals will you need? There is no stereotyped answer but rather a thorough case-by-case search to determine the adequate number required. For common significance testing, an investigator must determine a final sample size, and one difficult task is to precisely estimate this number before the data are collected. While statistical programs that include power and sample size analysis or input from a statistician can help in this matter, pilot studies may also be helpful to estimate such a number. It might also be based on a previous work or inspired by the scientific literature, although with some caution. The youngest of us might be impressed by the small number of animals used in published studies, but we should not be mistaken: unfortunately, more animals are often used than what is eventually reported in the literature. An obvious tendency to only show positive results seems to prevent authors reporting the actual numbers used. Despite our efforts to take the greatest care of all the subjects, each of us may occasionally lose an animal because of a shock in anesthesia or a postsurgical complication that we were unable to treat, a sudden death or another unknown reason. Feeling bad about it is normal, and this should never be considered as a personal failure. Scientific ethics and transparency should rather encourage us to share this kind of experience and to overtly integrate the potential loss of animals when planning an experiment.

Under the principle of reduction, the temptation is great to restrict the number of animals to the minimum for statistical purposes and, by excessive optimism, to ignore all events that may unexpectedly occur, further reduce this number, and ultimately compromise the validity of the study. Seeking a small number at all cost might be scientifically and ethically wrong because of the risk of a lack of statistical power (Button et al. 2013) and thus in contradiction with the 3R principle. On the opposite side of the spectrum, excessive cautiousness may incite setting the number to a much larger value than needed. There are, of course, experiments that demand a very high statistical power and therefore often require a very large sample size. Fortunately, the design and conduct of such studies may benefit from new statistical tools such as the variable-criteria sequential stopping rule (Fitts 2010a). This recently validated method allows the conduct of a study by stages, using a few subjects at a time, and to regularly determine if the planned experiment is worth pursuing or if the *p*-value is so high that it is unlikely that the expected significant effect will ever occur (Fitts 2010b).

The notion of reduction should thus be refined and a balance be found between the number of animals strictly necessary for the statistical demonstration and the number of animals needed to reach the final sample size. Should an inexperienced person perform the experiment, or if the brain target area is unknown to a research team, the total number of animals needed should include a small fraction for training and practice. Experimenters, even senior investigators, should not refrain from adding a practice or training subsection in their project. Far from showing a lack of competence, this is genuinely responsible and in the spirit of the 3R principle: increasing the number of animals now until your skills have adequately improved, or until you consistently reach the target area, so that you will not need as many later on.

1.2.4 Description of the Surgical Procedure

Like any other type of surgical procedure, stereotaxic surgeries in laboratory rodents should only be performed in healthy animals, and a thorough clinical examination before starting the procedure is a requisite. The protocol should therefore provide full details of the surgical procedure, starting with the preoperative examination of the health status of the subjects.

Before the surgery per se, all preoperative interventions should be indicated. Besides a physical exam, the experiment might require an overnight food/drink withdrawal, a complete blood count (CBC), or chemistry profile. Body temperature support (e.g., rectal probe and heating blanket), clipping of fur, and application of an ophthalmic ointment to the eyes are typical preoperative care interventions, applicable to both terminal and nonterminal surgeries. For a survival surgery, full details of aseptic techniques should also be given, along with the postoperative plan. Note that an exhaustive description of aseptic techniques should include details on the preparation of the surgical space, the surgeon(s) and the animal(s), and on the sterilization of instruments. Some institutions may ask for additional details on the area/zone where the surgery will be done. To comply with the principles of aseptic technique, such areas must be properly designed and maintained.

Details of the methods of induction and maintenance of anesthesia may often be presented in the form of a table (Table 1.1). Stereotaxic surgeries should not be performed in paralyzed animals without anesthesia. In such procedures, neuromuscular paralytics must always be combined with an adequate anesthetic agent. Experimenters are expected to use pharmaceutical-grade medications whenever available, even in acute procedures. Such medications are defined as drugs that can be purchased from a medical or veterinary supplier in a formulation that is ready to use in living vertebrate animals. It is usually not permitted to use non-pharmaceutical-grade compounds even in terminal studies, unless there is a scientific necessity or if acceptable veterinary or human pharmaceutical-grade products are not available. In such cases, a specific approval might be required and obtained from an IACUC or equivalent authority. In most countries, anesthetics, analgesics, and tranquilizing agents, such as ketamine or opioids, are controlled substances and may only be used under the handling and direction of a veterinary professional or principal investigator with an individual license. The license number and the products that the license covers should be indicated in the protocol form, either in the section listing the personnel qualifications or on a separate document, along with the list of persons authorized to handle the drugs under the license.

In addition to the dose, volume, route, and frequency of the drug administered, the description of the methods of anesthesia/analgesia should include the criteria used to assess and monitor their level/depth during the procedure and the recovery in case of a survival surgery. These criteria can include respiratory rate/effort, mucous membrane color, blood pressure, heart rate, body temperature, oxygen saturation, capillary refill time, electrocardiogram (EKG), and reflexes.

Table 1.1 Example of table describing the methods of induction and maintenance of anesthesia, copied from the 2012 version of the protocol form of the IACUC of the National University of Singapore

	Agent	Dose	Volume	Route	Frequency
Tranquilization					
Induction					
Maintenance					
Analgesia					

In accordance with the National Advisory Committee for Laboratory Animal Research of Singapore, animal use for research or teaching requires the submission of a protocol form available on the IACUC website http://www.nus.edu.sg/iacuc/

The description of the surgical procedure should include details on the intraoperative support, such as mechanical ventilation, heating to prevent hypothermia or cooling to prevent hyperthermia, and intravenous fluids. Should specimens be collected from the animals before, during, or after the surgery (blood, lymph, cerebrospinal fluid, gastrointestinal fluids, or solid tissues), their type, frequency, volume per collection, and the method of disposal should also be indicated.

The license form may contain a distinct section for the description of pain management throughout the experiment. Should the analgesia be withheld, a specific justification must be provided. Otherwise, this section may be completed with a list of the analgesics and dosing regimens.

1.2.5 Endpoints

There are two kinds of endpoints: experimental and humane. Experimental endpoints are defined as the point at which the scientific aims and objectives of the study have been reached. The humane endpoint is often defined as the point at which an animal's pain or distress is terminated, minimized, or reduced, by an action such as humane euthanasia, ending a painful procedure or administrating a treatment to relieve pain or distress. Whatever the type of study, e.g., regulatory or diagnostic toxicology and models of human diseases such as neurodegenerative studies or cancer, one must use the earliest endpoint that is compatible with the objective of the study (see also Appendices 3 and 4).

1.2.5.1 Experimental Endpoints

In the best-case scenario, the experiment on the animal is completed, and a decision is taken to euthanize the animal for histology. However, an unexpected event may incidentally occur during the experiment that will prevent any further observation on the animal. An example of such an event in stereotaxic surgeries is an unusual cranial suture scheme: after having checked the healthy status of an animal, having administered proper anesthesia/analgesia, and having placed it in the stereotaxic frame, the longitudinal incision of the skin may reveal cranial sutures that are not conform to the ones described in atlases. A decision has then to be taken as to whether the experiment should be pursued or not. Suture abnormalities can sometimes be overcome, for instance, by taking into account a slight (<200 μm) mediolateral misalignment of lambda and bregma sutures in the calculation of the coordinates for the target area (see details in Sect. 5.10). It is nevertheless worth considering interrupting the surgery at this point. Closing the wound and letting the animal gently recover (and possibly reuse it in another non-stereotaxic procedure) can be preferable to the pursuit of the procedure with the risk of missing the target area and unnecessarily using the animal.

Another unexpected event is the hemorrhagic disaster. Knowing the organization of the vessels at the surface of the brain, an angle of approach is typically applied to avoid rupturing a transverse sinus. Adjusting the position of the head and the coordinates of the craniotomy proves to be effective, and you are able to introduce a probe in the tenth rat of your series. Alas, after a few minutes and some hundreds of microns of advance of the probe, a dark red flow suddenly emerges along its shaft. With or without an action from the experimenter, the bleeding will generally stop at some point, but there is hardly a way to estimate the internal damage and the possible consequences of the hemorrhage. Here again, it will definitely be worth considering an early end of the experiment for the animal.

1.2.5.2 Humane Endpoints

The determination of the adequate humane endpoints requires a thorough observation of the animal and a scoring of its condition throughout the study. As mentioned above, surveillance checklists will provide a real advantage since they will allow an exhaustive examination of clinical signs that would otherwise be overlooked. When an endpoint is reached, a decision is taken to alleviate pain/distress either by administrating a treatment or by using an appropriate method for euthanasia. In the case of survival surgeries (e.g., injection of a molecule or compound for tract-tracing studies), criteria for an early interruption and euthanasia may include a rapid drop in body weight or a consistent and long period of weight loss that persists despite a change in the food regimen; prolonged lethargy; signs of a central nervous system disorder like tremors, spasms, convulsions, and paralysis; or clinical signs of an infectious disease.

When developing a new animal model or when the effects of a treatment are unknown, it can be very useful to conduct a preliminary or pilot study with a small number of animals in order to determine the morbidity, the time course of the effects, the most significant behavioral sign, and the frequency of observations that should be made in order to select the earliest endpoint. Another interest of such a study is to allow the collaborators in the study, particularly trainees and caretaker staff, to become familiar with the symptoms and signs expected from the procedure.

Table 1.2 Example of criteria for use on a scoring sheet. Scores from 0 to 3 can be attributed to each item and summed to establish a global score. Whenever an animal's score reaches a threshold defined prior to the experiment, a decision can be taken to euthanize the animal

Physical appearance	Body weight changes	Behavior	Other
0: normal	0: normal	0: normal	0
1: lack of grooming	1: <10 % weight loss	1: minor changes	1
2: rough coat, nasal/ocular discharge	2: 10–15 % weight loss	2: abnormal; reduced mobility, inactivity	2
3: very rough coat, abnormal posture	3: >15 % weight loss	3: lethargy, unsolicited vocalizations, self-mutilation	3

Once they have been listed, the different variables should be compiled in a scoring sheet, so that they can be monitored and scored at least once daily, preferably over a short time scale (Table 1.2). It will then be easy to set thresholds and define the corresponding actions to take (report to the veterinary staff, analgesic treatment, euthanasia) whenever the score of an animal reaches the predefined threshold. All persons participating in the study should receive appropriate guidance to fill in the scoring sheet.

1.2.5.3 Humane Euthanasia

Euthanasia is usually required at the end of a procedure that includes a stereotaxic surgery in order to check on histological brain sections the actual location of a target site. An appropriate method can be the one scheduled and approved in the protocol: the one specified in the establishment license or any method if the animal is already unconscious during a procedure (terminal surgery, e.g., non-survival neuronal recordings) and will not regain consciousness before death. Annex IV of the new Directive (2010/63/EU) enumerates the methods permitted for a humane killing of all species. For rodents, one of the following methods shall be used for euthanasia: anesthetic overdose (after prior sedation), carbon dioxide (except for rodent neonates) or inert gases (argon, nitrogen), cervical dislocation (after sedation for rodents above 150 g body weight), and concussion/percussive blow to the head. Decapitation should only be used if no other method is possible. Note that an accompanying method must be used to confirm the death (permanent cessation of heart and blood circulation, dislocation of the neck, exsanguination, onset of rigor mortis).

1.2.6 Compliance with the 3R Principle

In the 27 member states in Europe, experimenters are now required to assess the compliance of their project with the 3R principle. Useful help may be provided by project license forms that include a series of questions to answer. Below are those that are listed in the UK project license form.[14]

[14] The form can be found on the website of the Home Office, in the section on scientific development and research https://www.gov.uk/research-and-testing-using-animals, accessed September 10, 2013.

1.2.6.1 Replacement

Why is it not possible to achieve the objectives of your project without using animals?

What alternatives have you considered, and why are they not suitable? What alternatives will be used in achieving your objectives?

1.2.6.2 Reduction

What measures have been or will be taken to ensure that the minimum number of animals will be used in this project?

Explain the principles of experimental design you will use and any sources of advice you will consult, e.g., on statistics.

1.2.6.3 Refinement

Explain your choice of species, model(s), and method(s). Explain why they are the most refined for the intended purpose.

How will you minimize animal suffering in order to achieve your objectives?

Provide specific justification for any substantial severity protocols.

In Europe, the elements described in application of the three Rs should also be presented in nontechnical summaries (NTS). Publicly available, NTS are meant for the general public, not fellow scientists. When filing a project license form, applicants are therefore expected to avoid jargon and preferably use words that can be clearly understood by a large audience.

1.2.7 Duplication of Research

The duplication of harmful experiments in animals is in principle unacceptable. For this reason, and in addition to the compliance with the 3R principle, most legislative frameworks require that no unnecessary duplication of research be performed. "Duplication represents inadvertent, unconscious, or, more rarely, deliberate repetition of research efforts, thus not serving a need to confirm or otherwise verify conclusions from previous research undertakings" (Institute of Medicine and Committee on Co-Administration of Service and Research Programs of the National Institutes of Health 1991). This can apply to regulatory testing, for which each state/country is encouraged to recognize and accept published data from other states/countries, but research and teaching projects are concerned as well. Duplication should be distinguished from replication, the purpose of which is to deliberately repeat a previous study in order to confirm or extend still uncertain findings. Replication may involve repeating a study using the same methods. In the protocol approval process, experimenters should state in their own words why they feel that the procedure proposed for research or teaching is not unnecessarily duplicative. Such a statement must usually be accompanied by details on the literature/bibliographical searches (databases consulted, keywords used), as well as consultations with colleagues in the same field of specialty or other external experts (names and dates).

1.2.8 Responsibilities and Signature

Using animals in research is not trivial; it is actually a privilege granted by society to a restricted population of individuals, researchers, technicians, engineers, and students, with the expectation that this use will provide a significant benefit to knowledge or improvement in human and/or animal health and well-being. Using animals in research, testing, or teaching thus requires responsibility and sustained efforts to minimize pain and distress and promote their well-being. This is the reason why, in addition to national/federal laws, many institutions have implemented a fourth R to the 3R principle (Reduction, Replacement, Replacement, and responsibility).

Accordingly, a section of a license form is usually reserved for the principal investigator (PI) or project leader (PL), for assessment of responsibilities and signature. Whether or not it is explicitly written on the form, by signing the document, the PI/PL will attest that he/she has carefully considered the need to perform the procedure in live animals, the objectives of research or teaching, and the literature review and that he/she believes that the use of animals is the most appropriate method for attaining the teaching or research goal. The signature will also attest a personal engagement to ensure that animal use will follow the terms of the procedure described, to comply with policies; applicable federal, state, and local laws and regulations; ethical guidelines; and institutional and other policies.

1.2.9 Conclusion

Obtaining a license or approval from an IACUC for the duration of a project should not be considered as an end; by this means, a government or institution only acknowledges and records one's engagement to conduct a procedure responsibly. It would also be a mistake to believe that filing an animal protocol form to simply conform to regulations is acting in a morally responsible way. In the interactive process that will eventually lead to a license or approval, investigators are expected to be reactive and proactive, to show their genuine endeavor to adopt best practices, to develop even better ones, and to contribute to the global effort to improve the culture of care in laboratories. Getting an approval once should never interrupt the search for replacement or refinement alternatives that will not only improve animal welfare but also improve the quality of research findings.

References

Button KS, Ioannidis JP, Mokrysz C, Nosek BA, Flint J, Robinson ES, Munafo MR (2013) Power failure: why small sample size undermines the reliability of neuroscience. Nat Rev Neurosci 14:365–376

Canadian Council on Animal Care (2013a) Guide to the care and use of experimental animals : Chapter XIX (laboratory mice). http://www.ccac.ca/Documents/Standards/Guidelines/Vol2/mice.pdf. Accessed 10 Sept 2013

Canadian Council on Animal Care (2013b) Guide to the care and use of experimental animals: Chapter XXI (laboratory rats). http://www.ccac.ca/Documents/Standards/Guidelines/Vol2/rats.pdf. Accessed 10 Sept 2013

Committee for the Update of the Guide for the Care and Use of Laboratory Animals, Institute for Laboratory Animal Research, Division on Earth and Life Studies, National Research Council (2011) Guide for the care and use of laboratory animals, 8th edn. National Academies Press. ISBN 0309154006. http://www.nap.edu/catalog.php?record_id=12910. Accessed 10 Sept 2013

Committee on Guidelines for the Use of Animals in Neuroscience and Behavioral Research, National Research Council, Institute for Laboratory Animal Research, Division on Earth and Life Studies (2003) Guidelines for the care and use of mammals in neuroscience and behavioral research, National Academies Press. ISBN 0309089034. http://www.nap.edu/catalog.php?record_id=10732. Accessed 10 Sept 2013

Committee on Recognition and Alleviation of Pain in Laboratory Animals, Institute for Laboratory Animal Research, Division on Earth and Life Studies, National Research Council (2009) Recognition and alleviation of pain in laboratory animals. National Academies Press. ISBN 9780309128346. http://www.nap.edu/catalog.php?record_id=12526. Accessed 10 Sept 2013

Dubner R (1987) Research on pain mechanisms in animals. J Am Vet Med Assoc 191:1273–1276

Fitts DA (2010a) The variable-criteria sequential stopping rule: generality to unequal sample sizes, unequal variances, or to large ANOVAs. Behav Res Methods 42:918–929

Fitts DA (2010b) Improved stopping rules for the design of efficient small-sample experiments in biomedical and biobehavioral research. Behav Res Methods 42:3–22

Guillen J (2012) FELASA guidelines and recommendations. J Am Assoc Lab Anim Sci 51:311–321

Institute of Medicine, Committee on Co-Administration of Service and Research Programs of the National Institutes of Health, the Alcohol, Drug Abuse, and Mental Health Administration, and Related Agencies (1991) Research and service programs in the PHS: challenges in organization, National Academies Press, ISBN 9780309045810. http://download.nap.edu/cart/download.cgi?&record_id=1871. Accessed 10 Sept 2013

Kirik D, Rosenblad C, Bjorklund A (1998) Characterization of behavioral and neurodegenerative changes following partial lesions of the nigrostriatal dopamine system induced by intrastriatal 6-hydroxydopamine in the rat. Exp Neurol 152:259–277

Nicklas W, Baneux P, Boot R, Decelle T, Deeny AA, Fumanelli M, Illgen-Wilcke B (2002) Recommendations for the health monitoring of rodent and rabbit colonies in breeding and experimental units. Lab Anim 36:20–42

Office of Laboratory Animal Welfare, National Institute of Health (2002) Public health service policy on humane care and use of laboratory animals. http://grants.nih.gov/grants/olaw/references/phspol.htm. Accessed 10 Sept 2013

Rehbinder C, Baneux P, Forbes D, van Herck H, Nicklas W, Rugaya Z, Winkler G (1996) FELASA recommendations for the health monitoring of mouse, rat, hamster, gerbil, guinea pig and rabbit experimental units. Report of the Federation of European Laboratory Animal Science Associations (FELASA) Working Group on Animal Health accepted by the FELASA Board of Management, November 1995. Lab Anim 30:193–208

Russell WMS, Burch RL (1959) The principles of humane experimental technique. Methuen, London

Schwarting RK, Huston JP (1996a) Unilateral 6-hydroxydopamine lesions of meso-striatal dopamine neurons and their physiological sequelae. Prog Neurobiol 49:215–266

Schwarting RK, Huston JP (1996b) The unilateral 6-hydroxydopamine lesion model in behavioral brain research. Analysis of functional deficits, recovery and treatments. Prog Neurobiol 50:275–331

Ungerstedt U (1971) Postsynaptic supersensitivity after 6-hydroxy-dopamine induced degeneration of the nigro-striatal dopamine system. Acta Physiol Scand Suppl 367:69–93

Voipio HM, Baneux P, Gomez de Segura IA, Hau J, Wolfensohn S (2008) Guidelines for the veterinary care of laboratory animals: report of the FELASA/ECLAM/ESLAV Joint Working Group on Veterinary Care. Lab Anim 42:1–11

Chapter 2
Elements of Descriptive Neuroanatomy

2.1 General Organization of the Nervous System

The nervous system is the control center for communication between different organs in the body. The cells of the nervous system communicate via specific electrical signals that cause rapid responses by acting on muscle or glandular effectors. The nervous system receives sensory information in the form of stimuli processed by the sensory receptors and adapts the appropriate actions to take in response to this stimulation. The nervous system acts as the headquarters of management and the director of the response. When necessary, it can provide not only a motor reaction but also secretory responses through the endocrine system. The nervous system is composed of the central nervous system (CNS), which assures the integration and the regulation of the entire nervous system, and the peripheral nervous system (PNS), which allows communication between the CNS and the rest of the body.

2.1.1 The Central Nervous System (CNS)

During ontogenesis, the formation of the brain begins with the subdivision of the neural tube into three *primary* swellings or *vesicles* (Fig. 2.1a): at the front, the prosencephalon containing the visual and olfactive centers; in the middle, the mesencephalon; and, between the mesencephalon and the spinal cord, the rhombencephalon, containing the centers of auditory function. The prosencephalon divides further into two parts (Fig. 2.1b): the telencephalon, which will evolve into the two cerebral hemispheres, and the diencephalon wherein visual information is processed by the thalamus. The rhombencephalon subdivides into the metencephalon and myelencephalon (Fig. 2.2). Each of these five *secondary* embryonic vesicles generates the formation of highly specialized structures of the vertebrate nervous system.

B. Ferry et al., *Stereotaxic Neurosurgery in Laboratory Rodent*,
DOI 10.1007/978-2-8178-0472-9_2, © Springer-Verlag France 2014

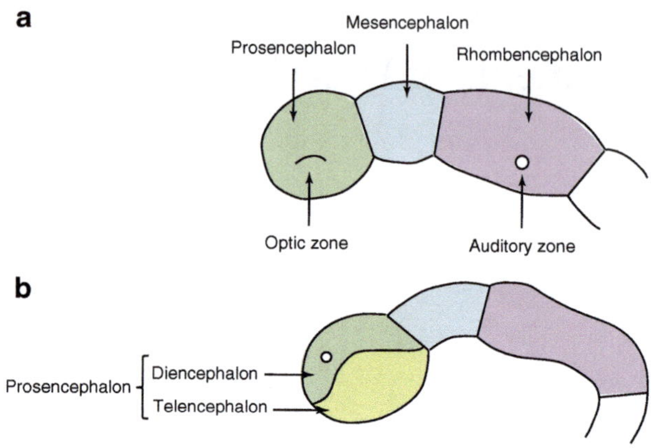

Fig. 2.1 Schematic representation of the three primary vesicles (**a**) and the subdivision of the prosencephalon (**b**)

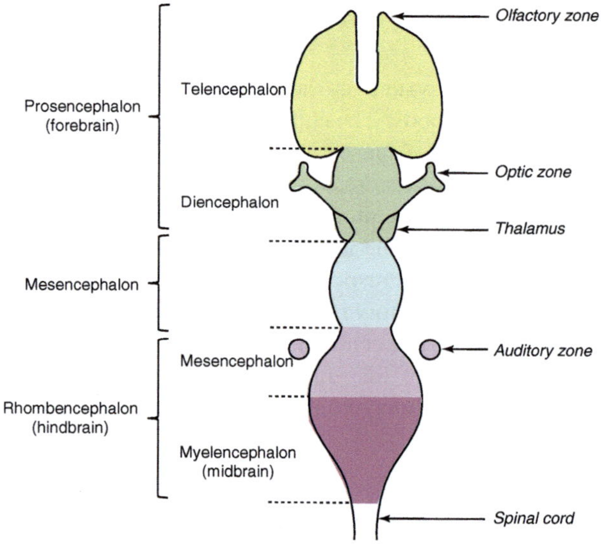

Fig. 2.2 Schematic representation of the five secondary vesicles, the main subdivisions of the embryonic vertebrate brain

After the formation of the two hemispheres, another pair of vesicles develops on their ventral surface to form the olfactory bulbs and other structures that contribute to olfaction. Diverse structures subsequently emerge from the walls of the telencephalon at the same time as the development of the white matter that connects them. The neurons of the telencephalic wall proliferate to form three regions: the cerebral cortex, the basal forebrain, and the olfactory bulbs (Saladin 2012; Mtui and Gruener 2006; Jessell et al. 2000; Duester 2008; Dono 2003). At this point in its evolution, the neuraxis reveals two principle and distinct elements: the brain situated within the cranial vault and the spinal cord within the vertebral canal (Table 2.1).

The brain is responsible for the emission of nervous impulses to the rest of the body, the processing and integration of information originating from sensory

Table 2.1 Anatomical divisions of the central nervous system

<table>
<tr><td rowspan="6">Central nervous system</td><td rowspan="5">Brain</td><td rowspan="2">Prosencephalon</td><td>Telencephalon</td><td colspan="2">Rhinencephalon, amygdala, hippocampus, neocortex, basal ganglia, lateral ventricles</td></tr>
<tr><td>Diencephalon</td><td colspan="2">Epithalamus, thalamus, hypothalamus, subthalamus, pituitary gland, pineal gland, third ventricle</td></tr>
<tr><td rowspan="3">Brain stem</td><td>Mesencephalon</td><td colspan="2">Tecum, cerebral peduncle, pretectum, mesencephalic duct</td></tr>
<tr><td rowspan="2">Rhombencephalon</td><td>Metencephalon</td><td>Pons, cerebellum</td></tr>
<tr><td>Myelencephalon</td><td>Medulla oblongata</td></tr>
<tr><td colspan="5">Spinal cord</td></tr>
</table>

nerves, and the more highly evolved operations that allow the association of this information so that the organism can respond appropriately. The brain is divided into four parts (Figs. 2.3, 2.4 and Table 2.1): the telencephalon or cerebrum, with its two large hemispheres (right and left) linked by the corpus callosum; the diencephalon containing the hypothalamus and thalamus; the mesencephalon or midbrain; and the rhombencephalon or hindbrain. The telencephalon and diencephalon together form the so-called forebrain. The mesencephalon (or midbrain) and rhombencephalon (or hindbrain, containing the pons and medulla oblongata) together make up the brain stem, which connects the rest of the brain to the cerebellum and spinal cord.

2.1.2 *Embryogenesis of the Principal Cerebral Structures* (Fig. 2.4)

The telencephalon designates an ensemble of structures that includes the cerebral hemispheres, which are subdivided into the cerebral cortex, the white matter, and the subcortical and associated structures. In phylogenetic terms, it is the most recent neural structure responsible for complex cognitive functions and is highly developed in mammals. The telencephalon is comprised of structures that are particularly implicated in olfaction, learning and memory, vigilance and attention, reward and reinforcement, pleasure and aversion, motivation, and higher affective functions.

The *diencephalon* underlies the two cerebral hemispheres and constitutes the structural center of the brain. It includes the thalamus, which is implicated in control of hunger and thirst, motivation, and sexual behavior. The dorsal thalamus controls the redistribution of sensory information, whereas the ventral thalamus has more of a motor role. The diencephalon is a center of integration and relay and of control of numerous functions essential for survival of the animal. It also includes the

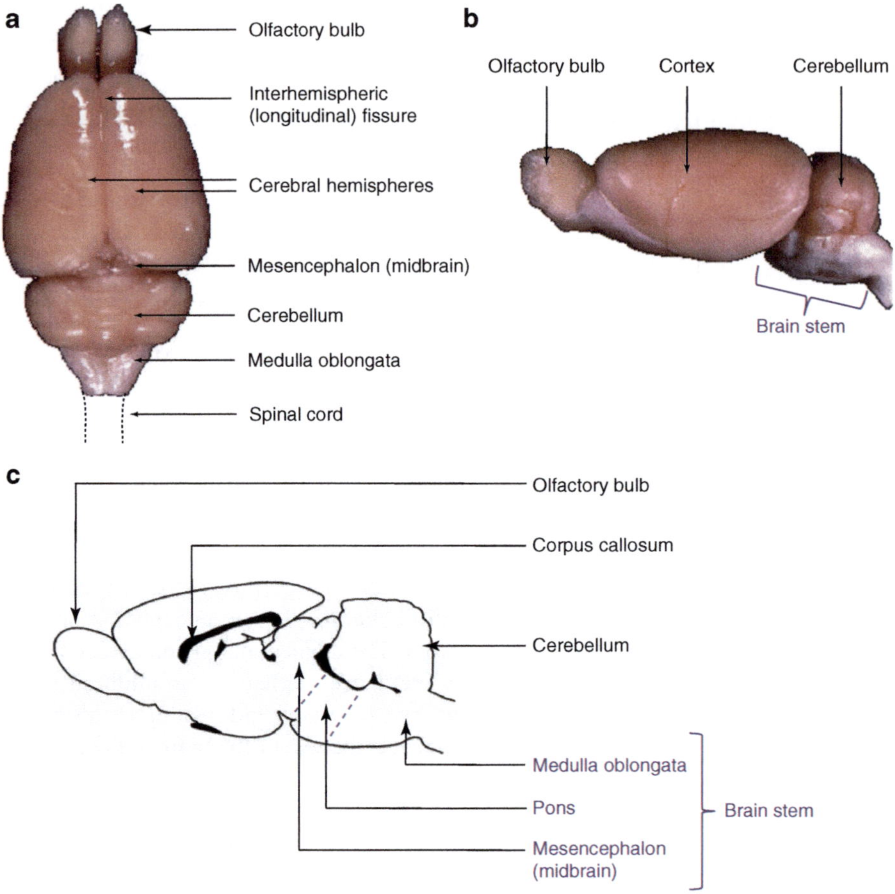

Fig. 2.3 Morphology of the rat brain. Principal subdivisions visible from a dorsal (**a**) and lateral (**b**) view and in sagittal section (**c**)

hypothalamus, which is composed of numerous small nuclei each possessing its own specific connections and neurochemical makeup. The hypothalamus is a higher center of integration of the autonomic and endocrine systems through the control that it exerts on the anterior pituitary. It is involved in major biological functions such as sleeping and waking, hunger and thirst, hormone secretion, and homeostatic functions such as blood pressure, temperature regulation, and fluid homeostasis. The epithalamus (pineal gland) regulates biological rhythms and skin pigmentation through the production and release of melatonin.

The *mesencephalon*, or midbrain, is a part of the brain stem composed of axon fiber bundles that are structurally and functionally connected to the rest of the brain via the cerebral peduncles, located between the pons and the diencephalon. Situated behind the peduncles are the tegmentum (the "floor" of the midbrain) and the colliculi, which are essential for visual and auditory information processing.

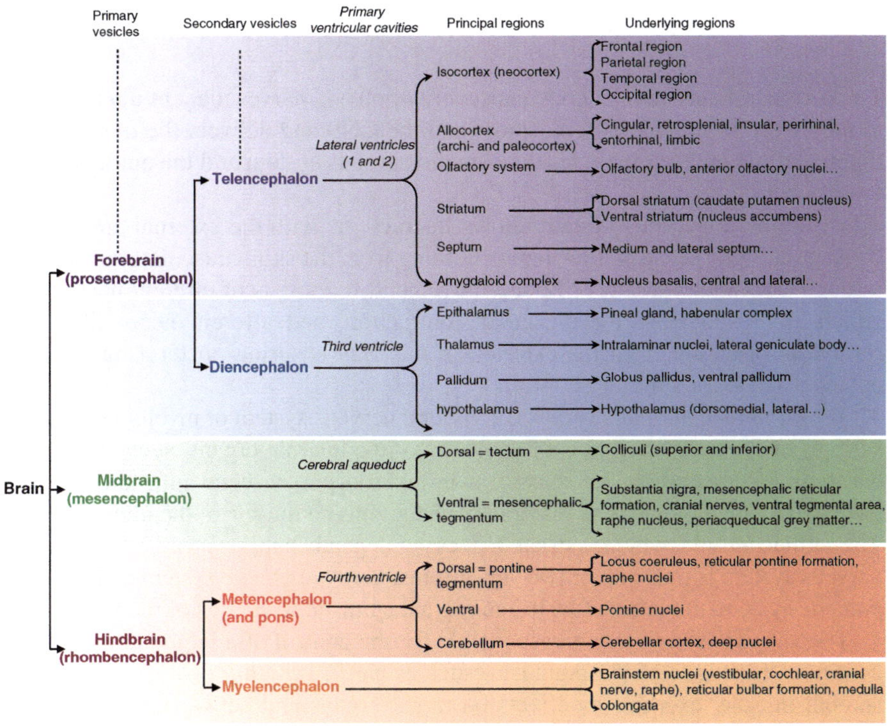

Fig. 2.4 Embryogenesis of the principal cerebral structures: from the three primary vesicles to the adult structures

The *metencephalon* is the center for coordination of muscular movements. It is composed of the neocortical cerebellum, in the form of two cerebellar hemispheres. The cerebellum is connected to the brain stem by three pairs of peduncles: the anterior peduncle, the posterior cerebellar peduncle, and the middle cerebellar peduncle. The latter, often referred to as the pons, is a complex structure. It plays a role in locomotor function by its position as a relay between the cerebellum and the rest of the brain and participates in autonomic system function and facial sensation. It is linked to the cerebellum by the middle cerebellar peduncle and makes up the anterior surface of the fourth ventricle. The cerebellar hemispheres are linked to the neocortex by nerve pathways that control complex voluntary movements.

The *myelencephalon* or medulla oblongata forms the posterior portion of the brain stem where autonomic reflex activities are situated. In vertebrates, it represents a transitional region between the spinal cord and the brain where ascending and descending sensory and motor information intersect. The myelencephalon houses most of the cranial nerves as well as centers that control respiration, cardiac rhythm, swallowing, coughing, vomiting, etc. It also contains an important gray matter structure, the olive, which appears in the form of an ovoid protuberance on its surface; the olive participates in motor control.

2.1.3 The Peripheral Nervous System

The peripheral nervous system consists of lymph and nerves that circulate information between motor-sensory receptors and the CNS and delivers the motor control outputs of the latter. It includes the somatic nervous system and the autonomic nervous system.

The somatic nervous system allows interactions with the external environment by responding to sensory information coming from the skin, musculature, and joints through a voluntary control of body movement. It consists of efferent nerve fibers, which are responsible for muscular contraction, and afferent nerve fibers that receive information from the exterior, in this way ensuring postural muscle tone, balance, and skilled motor activity.

The *autonomic nervous system* (or visceral nervous system or involuntary nervous system) innervates and directs the function of the internal organs, such as digestion and hormonal secretion. Its sensory neurons transport visceral information to the CNS and to the motoneurons that innervate the smooth muscle of the internal organs. The cardiac muscle, which is striated, does not depend on this system. The autonomic nervous system is made up of two subsystems: the *sympathetic* system and *parasympathetic* system, the latter often triggering antagonistic responses of the former.

The *sympathetic system* is responsible for the control of a large number of subconscious functions of the organism, such as cardiac rhythm or the contraction of smooth muscle, exerting its effects on target organs and cells principally via the catecholamine neurotransmitters noradrenaline and adrenaline.

The *parasympathetic system* controls involuntary functions of organs, glands, and blood vessels, conjointly with the sympathetic system. It uses acetylcholine as the principal neurotransmitter substance and is responsible for slowing cardiac frequency, increasing gastric secretions and increasing gastrointestinal tract motility. It is involved in certain pathological situations such as faintness, dizziness, colitis, and diarrhea.

2.1.4 The Meninges

The meninges are three membrane layers that envelope the CNS and serve as a physical and functional barrier between the nervous tissue and the surrounding bone (Fig. 2.5). The *dura mater* is a thick and resistant fibrous layer that strongly adheres to the bone and contains the venous sinuses. The arachnoid layer is a fine fiber network through which the cerebrospinal fluid can circulate; it sends protrusions, the arachnoid granulations (or villi), into the venous sinus of the dura mater. The *pia* is a thin and delicate fibrous tissue that is impermeable to fluid. This allows the pia mater to enclose the cerebrospinal fluid. By containing this fluid, the pia mater works with the other meningeal layers to protect and cushion the brain. It forms an intimate covering of the entire CNS but allows blood vessels to pass through and nourish the brain.

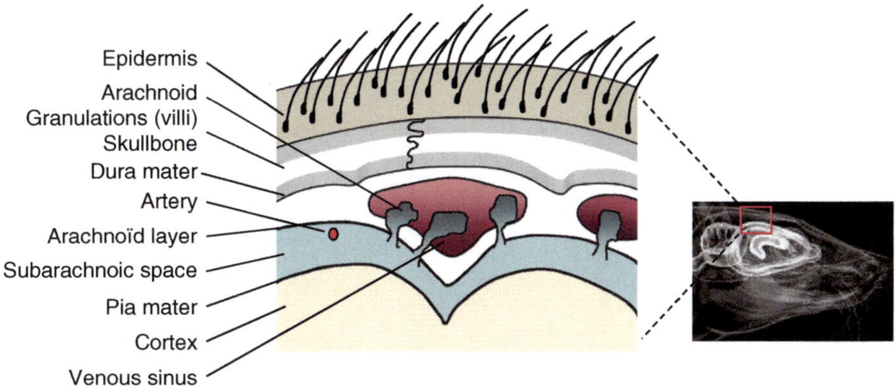

Fig. 2.5 The meninges and associated structures of the mammalian CNS. The figure on the *right* shows the position of a rat brain within its cranial cavity

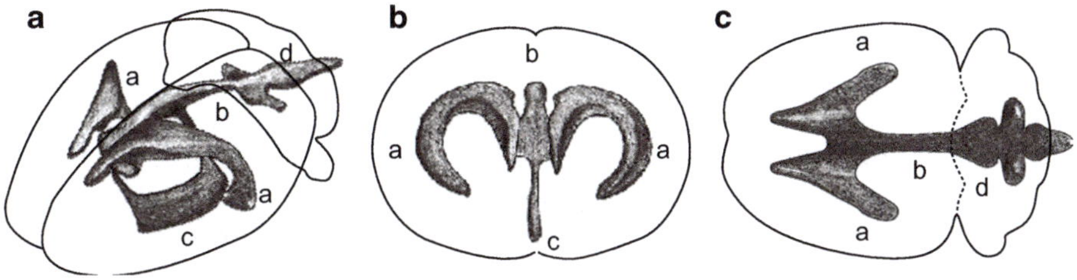

Fig. 2.6 The ventricular system (internal cavities) of the brain: (**a**) three-dimensional, (**b**) frontal, and (**c**) horizontal views. *a,* lateral ventricles; *b*, cerebral aqueduct; *c*, third ventricle; *d*, fourth ventricle

2.1.5 *Internal Cavities of the CNS: The Ventricular System*

The brain and spinal cord are continuously bathed with cerebrospinal fluid (CSF). The CSF occupies and circulates within these structures through a series of intercommunicating passages (the ventricular system) deep inside the brain and spinal cord and within the subarachnoid space between the arachnoid and pia mater (Fig. 2.6). CSF is a clear and colorless fluid produced by the choroid plexus in the lateral ventricles of the brain. It acts as a cushion or buffer for the cortex and the base of the brain, providing a basic mechanical and immunological protection to the brain, and serves a vital function in the autoregulation of cerebral blood flow. It circulates downward from the lateral ventricles through the foramina of Monro (interventricular foramen), third ventricle, aqueduct of Sylvius (cerebral aqueduct), fourth ventricle, foramen of Magendie (median aperture), and foramina of Luschka (lateral apertures) and through the subarachnoid space covering the entire surface of the brain and spinal cord. It is reabsorbed into the venous sinus blood via arachnoid granulations (villi) that provide for the elimination of metabolic and hormonal waste products.

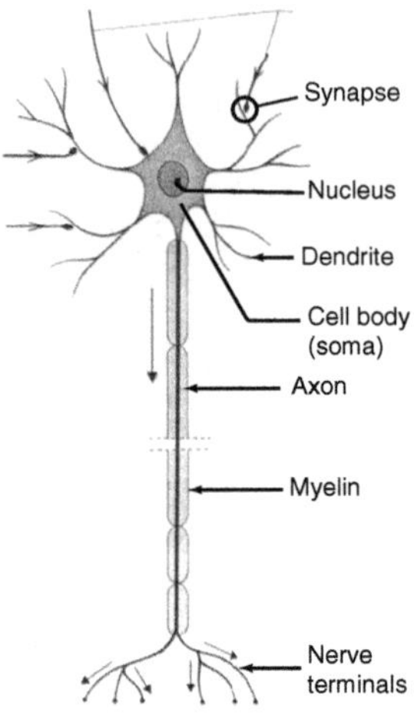

Fig. 2.7 Basic structure of a neuron

2.1.6 Nervous Tissue

The nervous tissue is made up of nerve cells or neurons and nonnervous or glial cells.

Examples of nerve cells are the pyramidal and granule cells (Fig. 2.7). The star-shaped astrocytes are an example of glial cells.

Neurons are distributed heterogeneously within the CNS, in zones with variable densities and organizations of cell bodies that make up the gray matter and in zones with relatively few or no cell bodies that form the white matter. Neurons can be classified according to the number of their neurites (axons and dendrites). Unipolar neurons have a single neurite, e.g., the rod and cone cells of the retina. Neurons with two neurites can be pseudo-unipolar (e.g., those in the spinal ganglia) or bipolar (e.g., in the vestibular ganglia). Neurons with more than two neurites, which are most common, are termed multipolar. Neurons are excitable cells, capable of generating and propagating bioelectrical signals called action potentials.

2.1.6.1 Glial Cells

Glial cells, also called neuroglia or simply glia (derived from the Greek "glue"), are nonneuronal cells that maintain homeostasis, form myelin, and provide support and protection for neurons in the brain and elsewhere, such as in the autonomic nervous system. The main functions of glial cells are to surround neurons and hold them in place, to supply nutrients and oxygen to neurons, to insulate one neuron from another, and to destroy pathogens and remove dead neurons.

Glial cells represent around 50 % of the cerebral mass. Two principal types of neuroglia exist: the macroglia that include astrocytes, oligodendrocytes, and Schwann cells and the microglia that include the macrophages. In contrast to the vast majority of neurons in the adult CNS, glial cells are capable of mitosis (division). Macroglial cells of the CNS and glia of the peripheral nervous system have different embryonic origins.

2.1.6.2 Macroglial Cells

Macroglia are derived from glioblasts of the embryonic tube neural. They include the *astrocytes*, the *oligodendrocytes*, and the *ependymocytes*.

Astrocytes are named as such because of their stellate form and multiple radiating prolongations. Astrocytes form a structural matrix for neurons and have a fundamental role in maintaining a balanced composition of the extracellular milieu. Feet-like extensions physically connect the astrocytes to the outside walls of capillaries when they are in close proximity to them. Through these "end feet," the astrocytes ensure a functional link between blood vessels and neurons: they take up energy substrates, glucose, and oxygen and transport them to the neurons. They also dispose of breakdown products from neurons for evacuation by the blood vessels. Astrocytes are the largest cells of the nervous tissue. Type I astrocytes are in close contact with blood capillaries. Type II astrocytes envelope neurons and synapses. Protoplasmic astrocytes are found in the gray matter, fibrous astrocytes in the white matter.

Oligodendrocytes are smaller than astrocytes and have fewer prolongations. They are the source of the myelin sheathing that covers nerve cell axons. One oligodendrocyte can myelinate an average of ten axons on various types of neurons in the CNS.

Ependymocytes (or ependymal cells) derive from ependymoblasts. They form a wall lining the ventricles of the CNS and are responsible for the synthesis of CSF at the level of the choroid plexus.

2.1.6.3 Glial Cells of the Peripheral Nervous System

Glial cells of the peripheral nervous system include satellite cells, Schwann cells, and microglia cells or microgliocytes. They are derived from the glioblasts of the embryonic neural crest. Satellite cells surround sensory and autonomic nerve cells and have a function analogous to that of the astrocytes. Schwann cells, like oligodendrocytes, provide for the myelinization of axons but form a myelin sheath around only a single axon.

The microglia, or microgliocytes, are macrophages that reside in the brain and spinal cord. They represent 5–25 % of all cells in the CNS and serve as the main active immune defense system for the CNS. They are morphologically and functionally different from the macroglial macrophages. Unlike macroglia that have a neuroectodermal origin, microglia cells are derived from blood cells. These small, stellate, and mobile cells are capable of phagocytosis and can proliferate in response to lesions in the CNS.

The implication of glial cells in the processing of nervous information was long ignored in comparison to the role played by neurons. Today it is recognized that glia can have important modulatory actions on neurotransmission and for this reason have now become a subject of intensive research.

2.2 Orientation and Reference Planes

A three-dimensional reference system is used to orient the CNS in space. Anatomical planes are defined according to three axes oriented perpendicularly to each other: a transverse axis (x), a vertical axis (y), and a horizontal axis (z) (Fig. 2.8). Each axis defines a specific plane that divides the brain into two parts (Fig. 2.9).

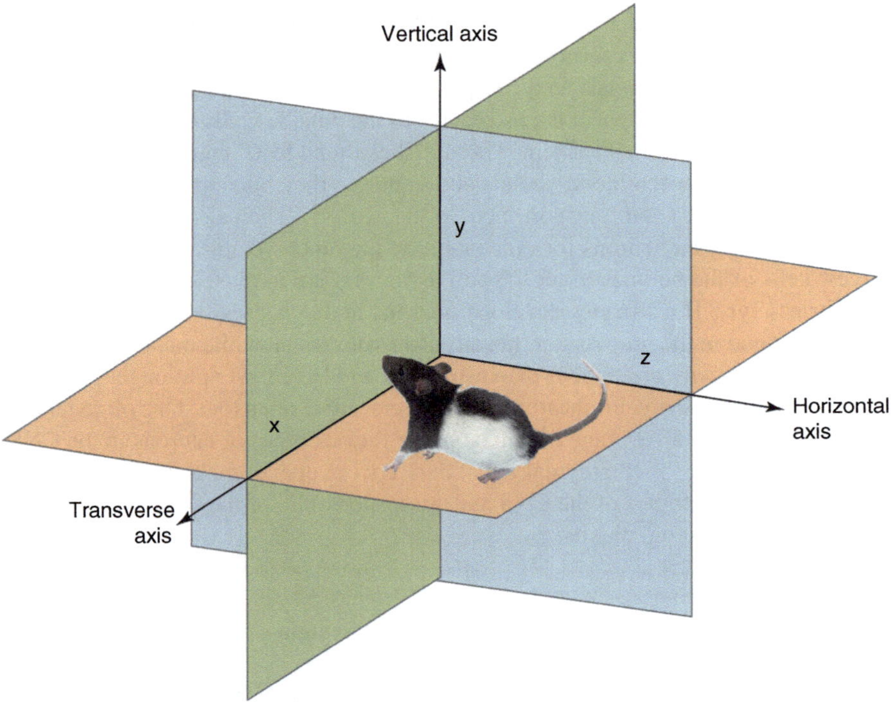

Fig. 2.8 The anatomical axes

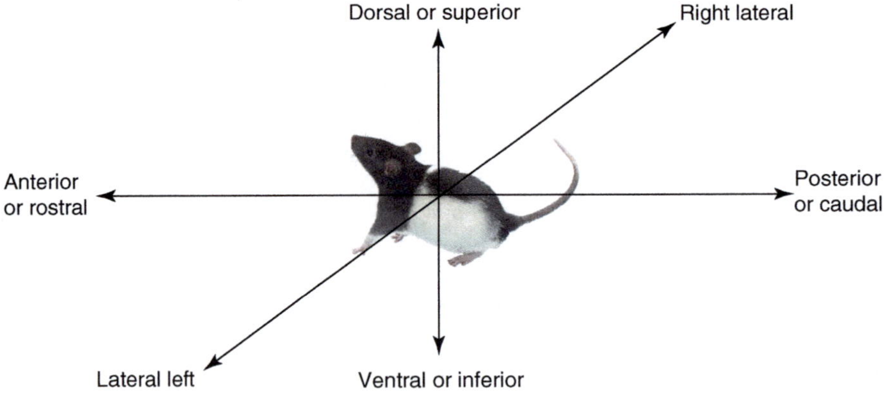

Fig. 2.9 The anatomical planes of orientation

The sagittal plane is a vertical plane that divides the brain into a left and right side, transecting the transverse axis (Fig. 2.10)

The transverse or frontal plane is a vertical plane perpendicular to the sagittal plane; it transects the horizontal axis and divides the brain into an anterior and a posterior part (Fig. 2.11).

The horizontal plane is parallel to the ground; it transects the vertical axis and divides the brain into an upper (dorsal) and a lower (ventral) portion (Fig. 2.12).

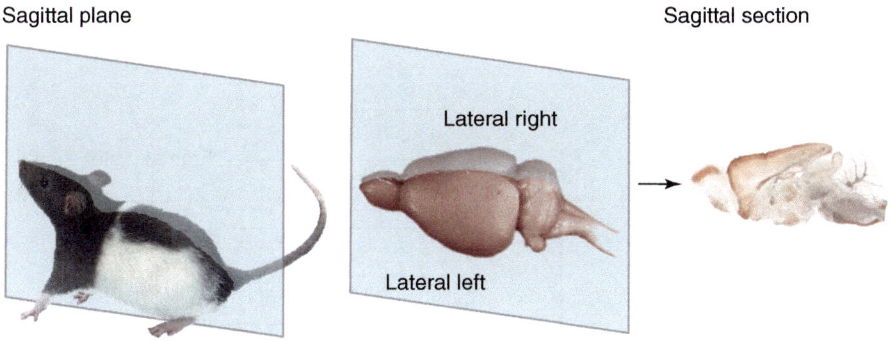

Fig. 2.10 Sagittal plane

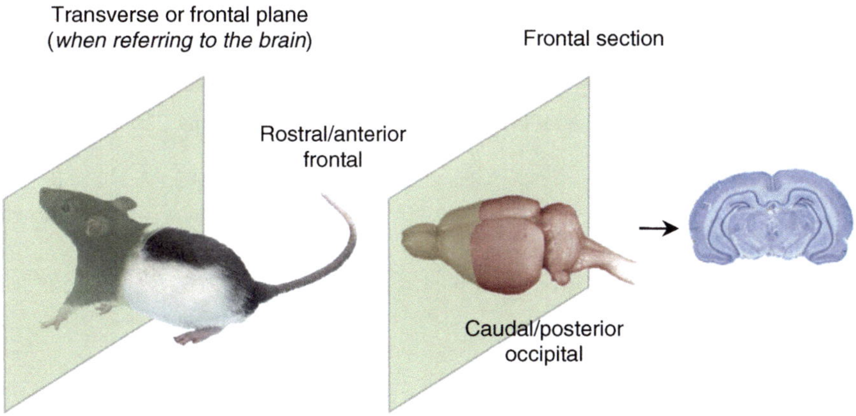

Fig. 2.11 Frontal plane

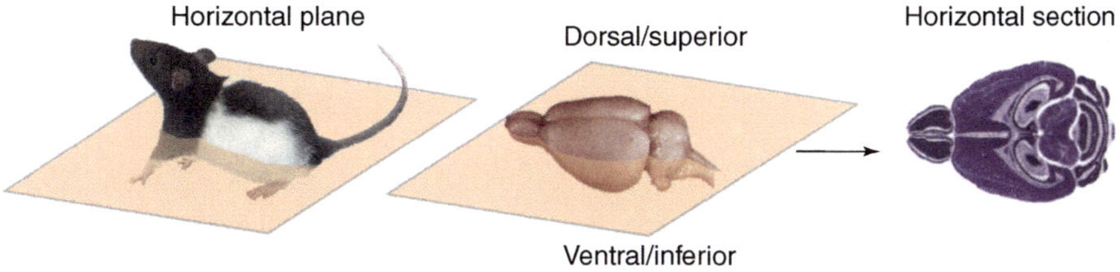

Fig. 2.12 Horizontal plane

2.3 Internal Morphology of the Rat Brain (from Paxinos and Watson 1998)

The reference point of bregma is defined in Chap. 3, paragraph 12. The following brain sections illustrate the topography of the principal brain structures.

2.3.1 Frontal Section: 4.70 mm Anterior to Bregma (Fig. 2.13)

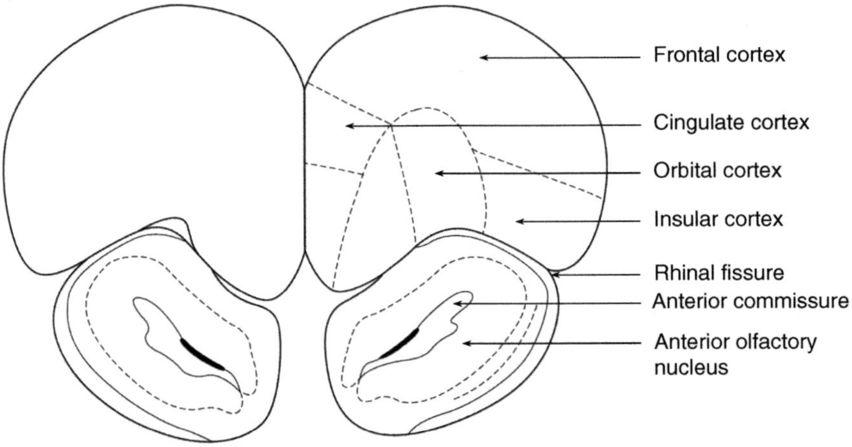

Fig. 2.13 Frontal section of the rat brain at 4.70 mm anterior to bregma (From plate n° 5 of Paxinos and Watson 1998)

2.3.2 Frontal Section: 1.20 mm Anterior to Bregma (Fig. 2.14)

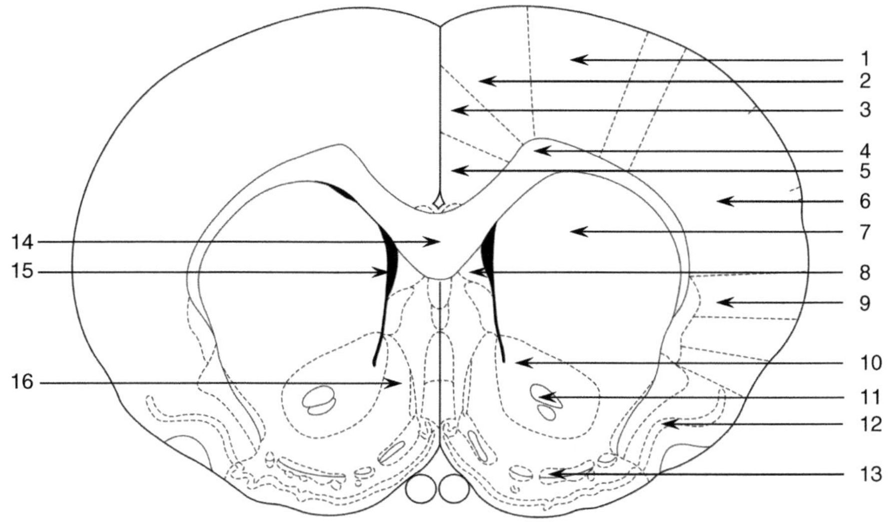

1: Primary motor cortex
2: Secondary motor cortex
3: Cingulate cortex 1
4: Cingulum
5: Cingulate cortex 2
6: Parietal cortex
7: Striatum
8: Lateral septum
9: Insular cortex
10: Noyau accumbens (core)
11: Anterior commissure
12: Pyriform cortex
13: Ventral pallidum
14: Corpus callosum
15: Lateral ventricle
16: Nucleus accumbens (shell)

Fig. 2.14 Frontal section of the rat brain at 1.20 mm anterior to bregma (From plate n° 13 of Paxinos and Watson 1998)

2.3.3 Frontal Section: 2.8 mm Posterior to Bregma (Fig. 2.15)

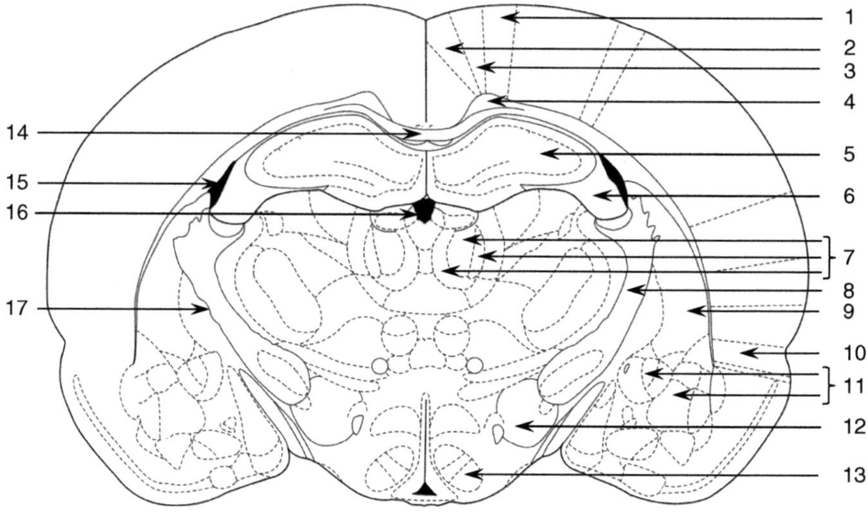

1: Primary motor cortex
2: Retrosplenial cortex
3: Secondary motor cortex
4: Cingulum
5: Hippocampal formation
6: Fimbria
7: Thalamus
8: Internal capsule
9: Striatum
10: Perirhinal cortex
11: Amygdaloid complex
12: Hypothalamus
13: Ventromedial hypothalamus
14: Corpus callosum
15: Lateral ventricle
16: Third ventricle
17: Globus pallidus

Fig. 2.15 Frontal section of the rat brain at 2.80 mm posterior to bregma (From plate n° 29 of Paxinos and Watson 1998)

2.3.4 Frontal Section: 6.30 mm Posterior to Bregma (Fig. 2.16)

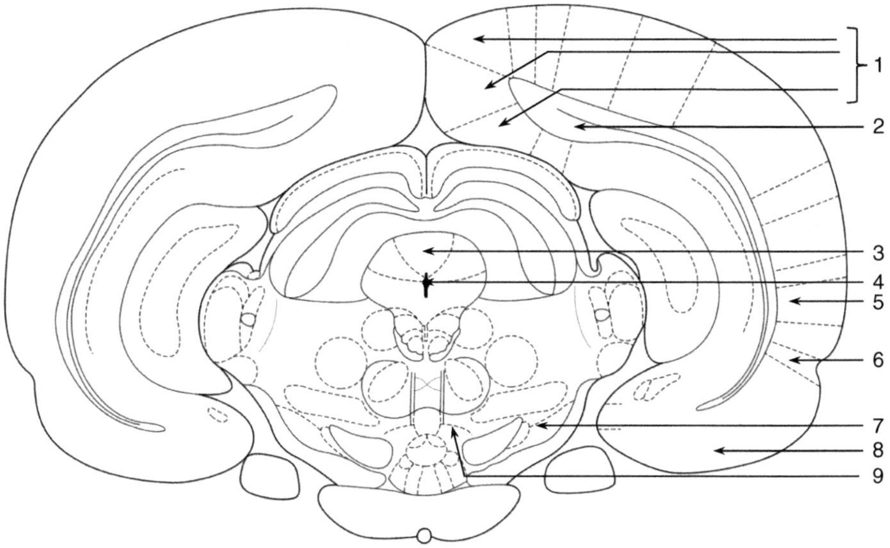

1: Retrosplenial cortex
2: Corpus callosum
3: Periacqueductal grey
4: Cerebral aqueduct
5: Temporal cortex
6: Perirhinal cortex
7: Substantia nigra
8: Entorhinal cortex
9: Ventral tegmental area

Fig. 2.16 Frontal section of the rat brain at 6.30 mm posterior to bregma (From plate n° 45 of Paxinos and Watson 1998)

2.3.5 Sagittal Section: 1.40 mm Lateral to Bregma (Fig. 2.17)

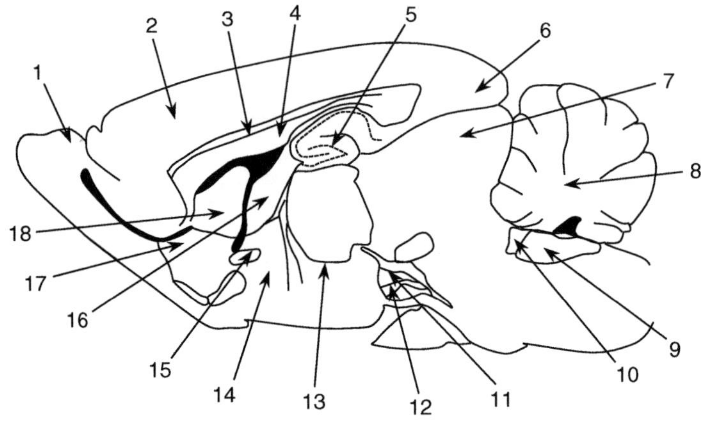

1: Olfactory bulb
2: Frontal cortex
3: Cingulum
4: Corpus callosum
5: Hippocampal formation
6: Occipital cortex
7: Mesencephalic (midbrain) tectum
8: Cerebellum
9: Vestibular nuclei
10: Locus coeruleus
11: Substantia nigra
12: Ventral tegmental area
13: Lateral hypothalamus
14: Medial forebrain bundle
15: Anterior commissure
16: Septum
17: Nucleus accumbens
18: Striatum

Fig. 2.17 Sagittal section of the rat brain at 1.4 mm lateral to midline (From plate n° 80 of Paxinos and Watson 1998)

2.3.6 *Some Useful References Concerning the Brain Anatomy*

Angevine J, Cotman C (1981) The principles of neuroanatomy. Oxford University Press, New York

Bloom F (1985) Brain, mind, and behavior. WH Freeman and Co. Press, New York

Butler A (1996) Comparative vertebrate neuroanatomy. Wiley-Liss, New York

Chiassen R, Radke W (1993) Laboratory anatomy of the vertebrates. WCB Publishers, Dubuque

Cohen M, Strumwasser F (1985) Comparative neurobiology. Wiley, New York

Eccles J (1973) The understanding of the brain. McGraw-Hill, New York

Karger S (1973) Central nervous system of vertebrates, vol 3, Part II. Arnold-Backlin-Strasse, New York

Kenneth S (2012) Anatomy and physiology: the unity of form and functions. McGraw Hill, New York

Kesner R, Olton D, Kesner R, Olton D (1990) Neurobiology of comparative cognition. Lawrence Erlbaum Associates, Hillsdale

Kuhlenbeck H (1973) Central nervous system of vertebrates, vol 3, Part II. Arnold-Backlin-Strasse, New York

Nichols J, Martin A, Wallace B (1992) From neuron to brain, 3rd edn. Sinauer Associates, Sunderland

Peters A, Palay S, Webster H (1991) The fine structure of the nervous system, 3rd edn. Oxford University Press, New York

Prosser C, Prosser C (1991) Neural and integrative animal physiology, 4th edn. Wiley-Liss, New York

Rice D, Barone S Jr (2000) Critical periods for the developing nervous system: evidence from humans and animal models. Environ Health Perspect 108(3): 511–533

Verley R (1977) Comparative development of the mammalian nervous system. Revue d'Électroencéphalographie et de Neurophysiologie Clinique 7(3):245–254

References

Dono R (2003) Fibroblast growth factors as regulators of central nervous system development and function. Am J Physiol Regul Integr Comp Physiol 284(4):867–881

Duester G (2008) Retinoic acid synthesis and signaling during early organogenesis. Cell 134(6):921–931

Jessell TM, Kandel ER, Schwartz JH (2000) Principles of neural science, 4th edn. McGraw-Hill, New York

Mtui E, Gruener G (2006) Clinical neuroanatomy and neuroscience. Saunders, Philadelphia

Paxinos G, Watson C (1998) The rat brain in stereotaxic coordinates, 4th edn. Academic, San Diego

Saladin K (2012) Anatomy & physiology: the unity of form and function, 6th edn. McGraw-Hill, New York

Chapter 3
Elements of Functional Neuroanatomy: The Major Neurotransmitter Systems

During the first half of the twentieth century, it seemed impossible to explain the rapidity of interneuronal signal transmission other than by the electric nature of the phenomenon. Experiments carried out in the 1920s by the German physiologist Otto Loewi showed that neurons could transmit electric signals to each other through chemical mediators. Since then, the concept of chemical support of communication between nervous cells has been universally accepted by neurobiologists. This information transfer is afforded by neurotransmitters, chemical substances that fulfill several strict criteria. They are synthesized within the neuron and stored within vesicles in the nerve terminal; they are released by exocytose in a calcium-dependent manner in response to the activation of the nerve terminal; the amount released must be necessary and sufficient to exert an action on the postsynaptic neuron or effector cell; after its release, the neurotransmitter binds to postsynaptic receptors and the remainder is eliminated by mechanisms of inactivation in the synaptic cleft; when the substance is administered exogenously at the appropriate concentrations, it must mimic the action of the endogenous released neurotransmitter. A large number of substances satisfy these criteria, but only a certain number of them will be addressed in this chapter: the amino acids glutamate, gamma-aminobutyric acid (GABA), and glycine and the monoamines noradrenaline, dopamine and serotonin, and acetylcholine (Bjorklund et al. 2005; Cooper et al. 2003; Epelbaum 1995; Meunier and Shvaloff 1995).

3.1 The Amino Acids

The majority of synapses in the brain utilize amino acids as neurotransmitters. The brain contains a large number of amino acids or their derivatives that induce excitation or inhibition when applied to neurons. The three principal amino acid transmitters are glutamate, gamma-aminobutyric acid (GABA), and glycine.

B. Ferry et al., *Stereotaxic Neurosurgery in Laboratory Rodent*,
DOI 10.1007/978-2-8178-0472-9_3,

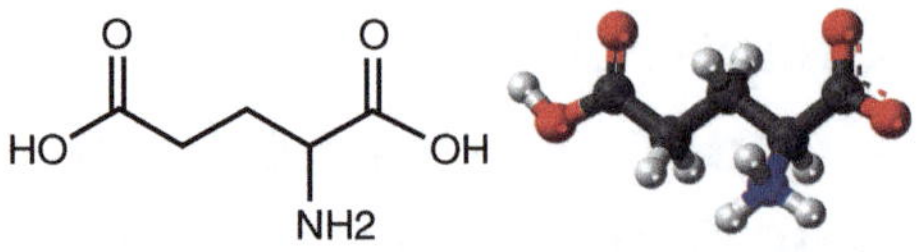

Fig. 3.1 Chemical formula and structure of glutamate

3.1.1 Glutamate

Glutamate (Fig. 3.1) is the principal excitatory neurotransmitter in the CNS. It is involved in phenomena of synaptic plasticity, such as long-term potentiation and trophic roles in development, but also in neuronal excitotoxicity related to excessive calcium influx.

3.1.1.1 Brain Localization

Neurons using glutamate as their neurotransmitter are among the most difficult to identify. In fact, in the brain, glutamate is an amino acid precursor in the synthesis of peptides and proteins, certain of which are neurotransmitters themselves, such as GABA. The identification of glutamatergic neurons is thus problematic in terms of distinguishing between their metabolic and neurotransmitter pools of glutamate. Glutamate is the principal transmitter in most of the efferent and intrinsic projections of the cerebral cortex (Fig. 3.2). It is also found in the thalamocortical projections, in the cerebellar cortex, in the hippocampal formation, etc.

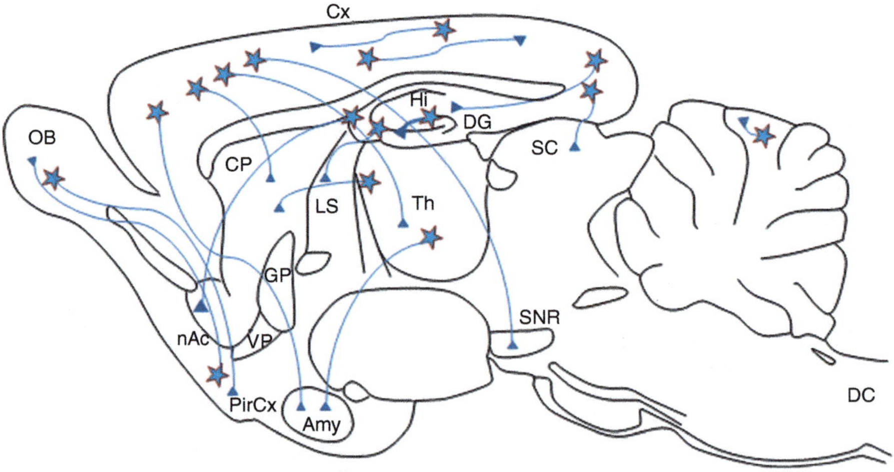

Fig. 3.2 Three-dimensional representation of glutamatergic pathways in the CNS. *nAc* nucleus accumbens, *Amy* amygdala, *OB* olfactory bulb, *DC* dorsal horn of the spinal cord, *CP* caudate-putamen, *SC* superior colliculus, *Cx* cerebral cortex, *PirCx* piriform cortex, *DG* dentate gyrus, *GP* globus pallidus, *Hi* hippocampus, *LS* lateral septum, *SNR* substantia nigra, *Th* thalamus, *VP* ventral pallidum

3.1.1.2 Biosynthesis and Degradation

Glutamate, or glutamic acid, is a nonessential amino acid synthesized by the organism and destined principally for protein synthesis. When utilized as a neurotransmitter, it is synthesized from glutamine within the axon terminal via the action of glutaminase, a mitochondrial enzyme (Fig. 3.6). The glutamate thus formed is stored in synaptic vesicles by an ATP-dependent/sodium-independent active transporter. In the case of glutamate neurons, glial cells recycle GABA and glutamine via the action of glutamine synthetase. Neurons and surrounding glial cells, including astrocytes, take up the glutamate released into the synaptic cleft via specific high-affinity transporters, thus inactivating glutamate's synaptic actions. After its reuptake, glutamate is converted by glutamine synthase to glutamine, which is then recycled to neurons for the synthesis of glutamate.

3.1.1.3 Glutamate Receptors

Glutamate receptors (see Fig. 3.3) are classified into ionotropic and metabotropic receptors.

Receptors		Agonists	Antagonists
NMDA	Glutamate binding sites	Trans-ACBD	AP-5 CPP
	Glycine binding sites	Glycine	7-Chlorokynurenic acid
	Ion channel		Phencyclidine (PCP) MK-801
AMPA		AMPA quisqualate	CNQX DNQX GYK153655
Kainate		Kainate Domoic acid ATPA	LY466195 LY382884 EAB-318
Metabotropic	Group I	Quisqualate	AIDA
	Group II	L-CCG-I	LY 341495
	Group III	L-AP4	CPPG

Fig. 3.3 Glutamate receptors and their agonist and antagonist ligands (Danysz and Parsons 1998; Berger et al. 1994; Parsons 2001; Yang et al. 2008; Weiss et al. 2006; Sun et al. 2004; Binns et al. 2003)

Ionotropic or Ion Channel Receptors

Three types of ionotropic receptor are subclassified according to the exogenous or endogenous agonist substances that they selectively bind:

- NMDA receptors are made up of a transmembrane protein complex containing several binding sites and an ion channel permeable to calcium, sodium, and potassium ions. The principal binding site permits the interaction of agonists such as glutamate, NMDA itself, and competitive antagonists such as D-2-amino-5-phosphono-pentanoic acid (D-AP5) and 3-[(+)2-carboxypiperazine- 4-yl] propyl-1-phosphonic acid (CPP). An allosteric site binds glycine and potentiates the action of glutamate. Glycine is indispensable for the activation of the receptor and is blocked competitively by 7-chloro-kynurenate. Two sites are situated within the ion channel: a voltage-dependent cationic site that binds free magnesium (Mg^{2+}) and blocks the calcium channel and a site which binds synthetic compounds such as certain psychotomimetic drugs that act as noncompetitive NMDA receptor antagonists, e.g., phencyclidine (PCP), ketamine, and [5R,10S]-[+]-5-methyl-10,11-dihydro-5*H*-dibenzo[*a*,*d*]cyclohepten-5,10-imine (MK-801, dizocilpine).
- AMPA (alpha-amino-3-hydroxy-5-méthyl-4-isoxazole-propionic acid) receptors are ion channels permeable to monovalent cations such as sodium and potassium. Agonists include AMPA itself, quisqualate, and glutamate. There is no specific antagonist for the receptor but certain compounds show a relative selectivity, including 6-cyano-7-nitroquinoxaline-2,3-dione (CNQX) and 6,7-dinitroquinoxaline-2,3-dione (DNQX).
- Kainate receptors are permeable to sodium, potassium, and calcium. Agonists include kainic acid, domoic acid, and glutamate. DNQX has been used as a non-selective antagonist.

Metabotropic Receptors

Eight different types of metabotropic glutamate receptors (mGluR1 to mGluR8) are divided into three groups, based on receptor structure and physiological activity. Group I is comprised of receptors mGluR1 and mGluR5. Their activation stimulates the inositol phosphate/diacylglycerol pathway via a pertussis-toxin-sensitive G protein. This induces a rapid increase of intracellular calcium and the activation of protein kinase C. Group II includes mGluR2 and mGluR3 receptors, while group III is made up of GluR4, mGluR6, mGluR7, and mGluR8 receptors. The receptors of these two latter groups inhibit adenylate cyclase via a Gi protein. Selective agonists and antagonists exist for the different receptor groups (see Fig. 3.3).

3.1.1.4 Implication of Glutamate in CNS Disorders

The prolonged application of high concentrations of glutamate has neurotoxic effects resulting from a massive influx of calcium into the neuron – this phenomenon is referred to as excitotoxicity. The neurotoxic property of glutamate and related substances plays an important role in neuronal lesions associated with epilepsy, cerebral ischemia, and traumatic brain injury. Since NMDA antagonist substances can provoke psychotic symptomatologies, it has been suggested that certain psychiatric diseases, such as schizophrenia, may be related to abnormalities in the glutamatergic system.

3.1.1.5 Experimental Tools Influencing the Glutamatergic System

A large number of pharmacological agents are capable of modulating glutamate receptor function (see Fig. 3.3). A selective neurotoxin for the glutamatergic system does not exist. However, it is possible to produce excitotoxic lesions of neurons that express glutamatergic receptors, either directly with glutamate receptor agonists such as kainic acid, ibotenic acid, or NMDA or indirectly, for example, by experimentally provoking an ischemia.

3.1.2 Gamma-Aminobutyric Acid (GABA)

GABA is the major inhibitory neurotransmitter in the nervous system (Fig. 3.4). Its importance is underlined by its deficits associated with epilepsy and anxiety.

3.1.2.1 Brain Localization

GABA is derived from glutamate and is specific to nervous tissue. The cartography of the GABAergic systems (Fig. 3.5) can be revealed using antibodies specific for

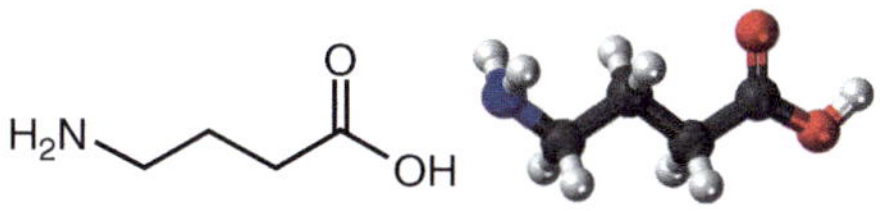

Fig. 3.4 Chemical formula and structure of GABA

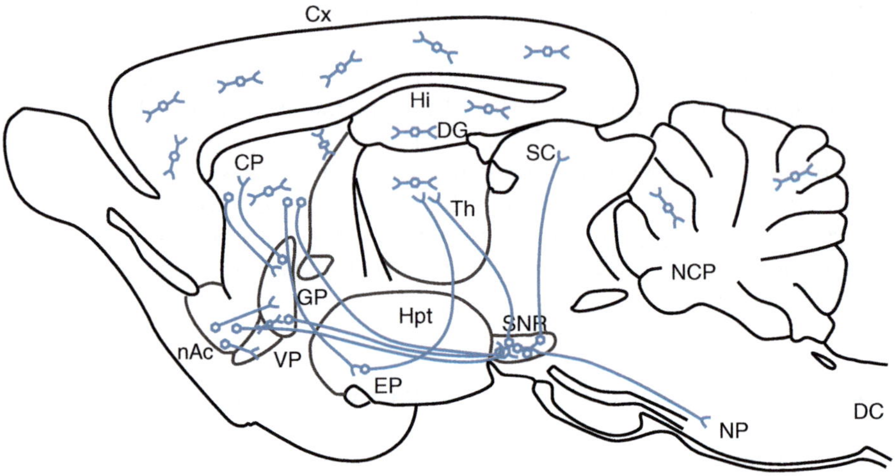

Fig. 3.5 Three-dimensional representation of GABAergic pathways in the CNS (From Epelbaum 1995). *nAc* nucleus accumbens, *CP* caudate-putamen, *SC* superior colliculus, *Cx* cerebral cortex, *DC* dorsal horn of the spinal cord, *EP* entopeduncular nucleus, *DG* dentate gyrus, *GP* globus pallidus, *Hi* hippocampus, *Hpt* hypothalamus, *NCP* deep cerebellar nuclei, *NP* pontine nuclei, *SNR* substantia nigra reticulata, *Th* thalamus, *VP* ventral pallidum

either GABA or its enzyme of synthesis, glutamate decarboxylase (GAD). GABAergic interneurons are distributed throughout the CNS, including the olfactory bulb, cortex, and hippocampus. Very few structures in the rat brain, such as the ventroposterior thalamic complex, are devoid of these interneurons. The major GABAergic projection systems include the striato-pallido-thalamic, the striato-nigro-thalamic, and the striato-nigro-tectal pathways.

3.1.2.2 **Biosynthesis and Degradation** (Fig. 3.6)

GAD catalyzes the synthesis of GABA by the decarboxylation of glutamate derived from glutamine. This step requires the presence of the cofactor, pyridoxyl phosphate (PLP or vitamin B6). After its release into the synaptic cleft, GABA is rapidly inactivated by high-affinity reuptake transport systems in both neurons and glia. It is subsequently converted to succinate by two mitochondrial enzymes, GABA transaminase and succinate-semialdehyde dehydrogenase. Succinate thus formed then enters the Krebs cycle. A portion of cellular GABA is not utilized as transmitter – after uptake into mitochondria by membrane exchange, it participates in oxidative metabolism.

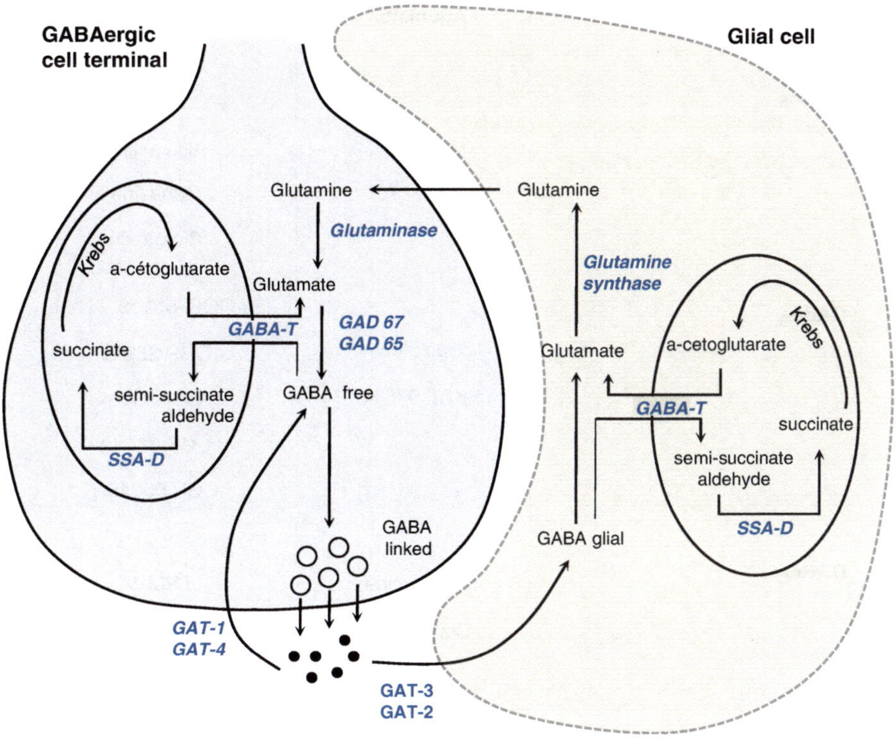

Fig. 3.6 Pathways of biosynthesis and degradation of GABA (From Epelbaum 1995). *GABA-T* GABA transaminase, *GAD* glutamate decarboxylase, *GAT* GABA transporter, *SSA-D* succinate-semialdehyde dehydrogenase

3.1.2.3 GABA Receptors

Three major types of GABAergic receptors exist: $GABA_A$, $GABA_B$, and $GABA_C$ (Fig. 3.7).

- $GABA_A$ receptors (ionotropic) are coupled to an ion channel specific for chloride ions, the entry of which leads to a hyperpolarization of the cell. This receptor has several different types of binding sites – a site for GABA itself; a benzodiazepine site that potentiates the action of bound GABA through an allosteric effect; a third site that can indiscriminately bind barbiturates, steroids, and alcohol; and a fourth site that binds picrotoxin, a noncompetitive antagonist of GABA. The principal competitive ligands for the $GABA_A$ receptor include the agonist muscimol and the antagonist bicuculline (Labrakakis et al. 2003; Bolea et al. 1999; Mirza et al. 2008).

Receptors	Agonists	Antagonists
$GABA_A$	Muscimol NS11394	SR95531 Bicuculline Gabazine Picrotoxin
$GABA_B$	Baclofen SK&F 97541	CGP–56119 CGP 35348 Phaclofen CGP 36742 CGP 55845A
$GABA_C$	Isoguvacine Cis-4-ACA	TPMPA SGS742 CGP36742 Cyclothiazide Saclofen

Fig. 3.7 GABA receptors and their agonist and antagonist ligands (Labrakakis et al. 2003; Bolea et al. 1999; Funahashi and Stewart 1998; Shen and Johnson 1997; Kabashima et al. 1997; Patenaude et al. 2003; Mirza et al. 2008; Bon and Galvan 1996; Chebib et al. 2009; Xie et al. 2008; Reis and Duarte 2006; Yang et al. 2003)

- $GABA_B$ receptors (metabotropic) are coupled to a *Go* or *Gi* protein. The *Go* protein can be coupled to a calcium channel or a potassium channel, while the *Gi* protein inhibits adenylate cyclase. The activation of these receptors produces a decrease in calcium currents and as a result reduces the release of neurotransmitters at the nerve terminal level. Their activation also increases potassium ion conductance that causes a hyperpolarization of postsynaptic neurons and a reduction of cAMP leading to reduced transmitter release. Principal ligands for this receptor include the agonist baclofen and the antagonist CGP 56119 (Patenaude et al. 2003).
- The $GABA_C$ receptor (ionotropic) is coupled to an ion channel specific for chloride ions. Principal ligands include the agonist isoguvacine and the antagonist TPMPA (Yang et al. 2003; Xie et al. 2008).

3.1.2.4 Implication of GABA in CNS Disorders

A deficit in GABAergic transmission is potentially epileptogenic. Conversely, an increase in GABAergic transmission can have sedative effects as seen with the use of benzodiazepines and barbiturates. An implication of the GABAergic system in certain anxiety disorders seems likely. An alteration in cerebral GABA metabolism has also been associated with neurodegenerative disorders such as Huntington's disease, in which striatonigral GABAergic neurons are affected.

3.1.2.5 Experimental Tools Influencing the GABAergic System

GABAergic transmission can be modified at different levels: synthesis, degradation, and reuptake. Inhibitors of GAD, such as allylglycine and cycloserine, block GABA synthesis. GABA transaminase, the enzyme of degradation, is inhibited by gamma-vinyl-GABA. Succinate-semialdehyde dehydrogenase is inhibited by sodium valproate. GABA reuptake is inhibited by derivatives of nipecotic acid. Numerous pharmacological agents can modify GABA receptor function (see Fig. 3.7). However, neurotoxins selective for GABAergic neurons are nonexistent.

3.2 The Monoamines

3.2.1 Catecholamines Derived from Tyrosine

3.2.1.1 Noradrenaline

Noradrenaline (Fig. 3.8) has an anatomical and functional duality, as a hormone synthesized and secreted by the adrenal gland and as a neurotransmitter in the central and peripheral nervous systems. Noradrenaline is probably the transmitter

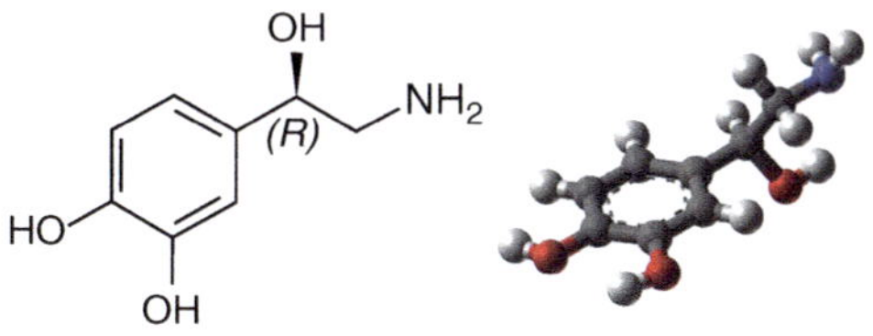

Fig. 3.8 Chemical formula and structure of noradrenaline

implicated in the widest diversity of functions, from the control of blood pressure, attention, and the sleep-wake cycle to cerebral synaptic plasticity.

Brain Localization

Histochemical techniques have permitted the localization of noradrenergic neurons by the revelation of noradrenaline and its rate-limiting enzyme of synthesis, dopamine-beta-hydroxylase (DBH). Noradrenergic nerve cell bodies in the CNS are concentrated within the brain stem, in the region bridging the locus coeruleus; the cell groups A4, A5, A6, and A7; the lateral reticular nuclei of the medulla oblongata (group A1); and the nucleus of the tractus solitarius (group A2) (Fig. 3.9; Dahlstrom and Fuxe 1964). Axon bundles from these nuclei disperse widely throughout the CNS to innervate almost every part of the brain. The locus coeruleus has the most diffuse connections in the brain, a single neuron from this nucleus forming as many as 25,000 synapses. Because of this ubiquitous distribution, the locus coeruleus can influence virtually the entire brain.

Biosynthesis and Degradation

Noradrenaline is synthesized from the amino acid tyrosine (Fig. 3.10). The active uptake of tyrosine by catecholaminergic neurons is followed by its hydroxylation to L-dihydroxy-phenylalanine (L-Dopa) by tyrosine hydroxylase. L-Dopa is then decarboxylated to form dopamine, which is then taken up and stored within synaptic vesicles where it is converted to noradrenaline by dopamine-beta-hydroxylase.

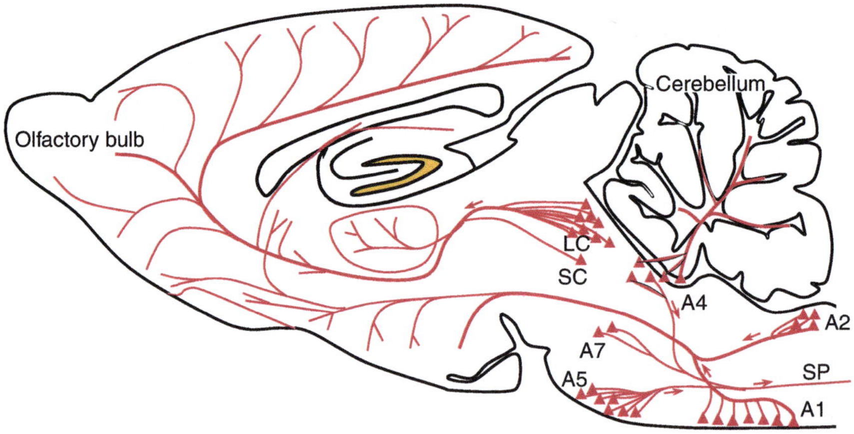

Fig. 3.9 Noradrenergic neurons and their principal projections. Zones A1, A2, A4, A5, and A7 contain noradrenergic neurons. *LC* locus coeruleus, *SC* subcoeruleus area, *SP* spinal cord, *LC* represents the zone A6

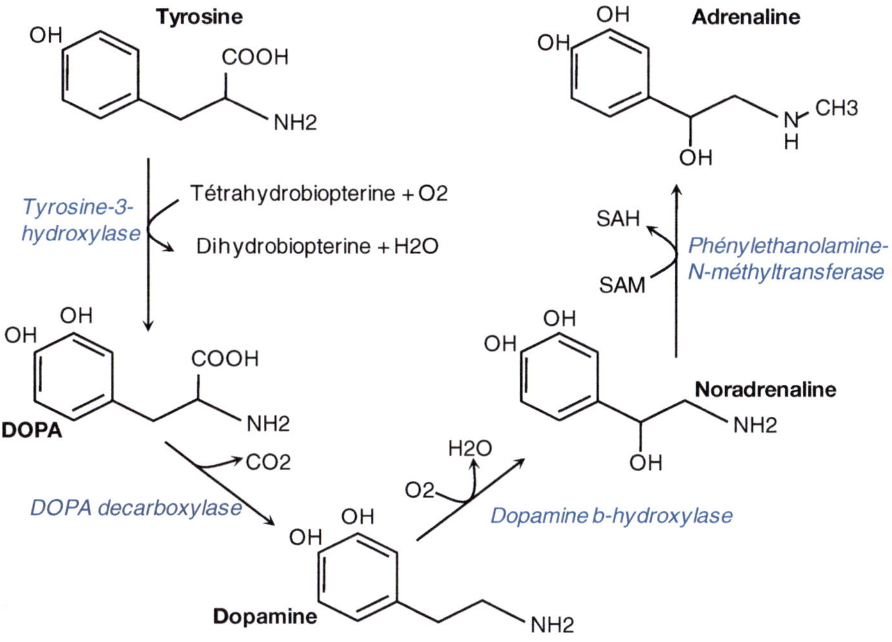

Fig. 3.10 Biosynthetic pathways of catecholamines (From Meunier and Shvaloff 1995)

Released noradrenaline is inactivated by several mechanisms (see Fig. 3.11). A minor portion in the synaptic cleft is transformed directly to normetanephrine by the enzyme catechol-*O*-methyltransferase (COMT) located in surrounding non-catecholaminergic neurons and glia. The majority of released noradrenaline is taken back up into the nerve terminals by a transmembrane transport system that is specific for noradrenaline. Within the nerve terminal, the recaptured noradrenaline is metabolized to 3,4-dihydroxy-phenyl-glycol-aldehyde by monoamine oxidase A (MAO-A), which converts the functional amine group to an aldehyde. The aldehyde is subsequently transformed to the alcohol and carboxylic acid derivatives.

Receptors

Two types of adrenergic receptors exist, alpha and beta, which are divided into the following subtypes: $alpha_{1a}$, $alpha_{1b}$, $alpha_{1d}$, $alpha_{2a}$, $alpha_{2b}$, $alpha_{2c}$, $beta_1$, $beta_2$, and $beta_3$ (Fig. 3.12; Gether et al. 2002). The stimulation of $alpha_1$ receptors activates the metabolism of phosphoinositides via a *Gq* protein. These receptors are located for the most part on postsynaptic sites. $Alpha_2$ receptors are coupled to *Gi* or *Go* proteins and their activation leads to an inhibition of adenylate cyclase; they are principally localized to presynaptic sites on noradrenergic neurons where they play a role in the negative feedback control of noradrenaline synthesis and release. $Alpha_2$ receptors have also been identified on glutamatergic, serotoninergic, and

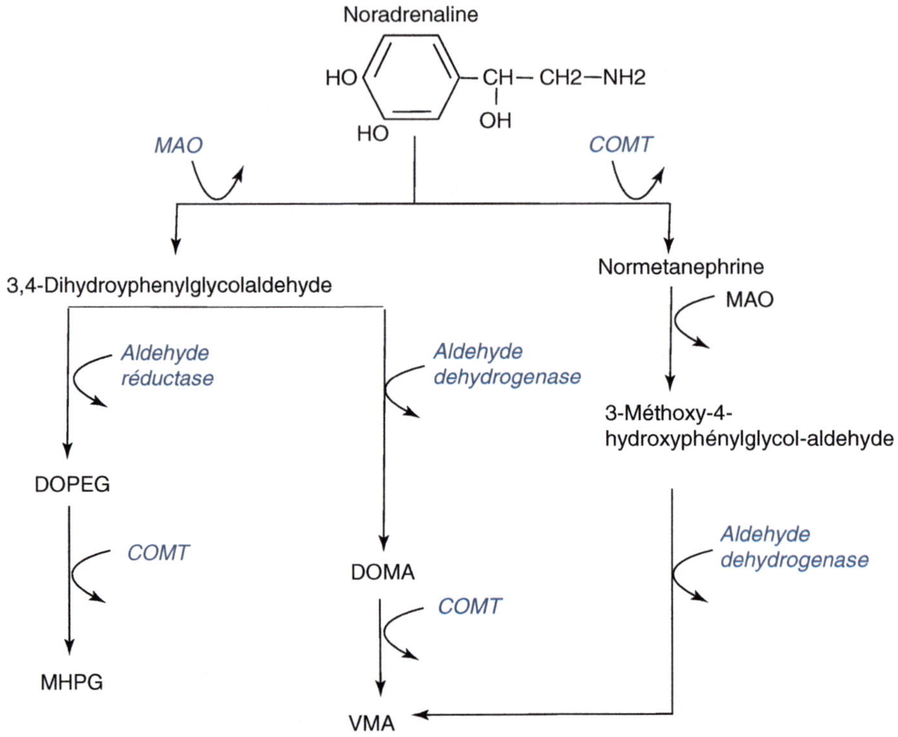

Fig. 3.11 Pathways of noradrenaline degradation. *DOPEG* dihydroxy-phenyl-ethylene-glycol, *MAO* monoamine oxidase, *COMT* catechol-*O*-methyltransferase, *MHPG* 3-methoxy-4-hydroxy-phenyl-ethylene-glycol, *DOMA* dihydroxymandelic acid, *VMA* vanillylmandelic acid

cholinergic neurons. Beta-adrenergic receptors are localized to postsynaptic sites where their function is excitatory or facilitatory; they are principally coupled to a *Gs* protein and their stimulation leads to an activation of adenylate cyclase and the formation of cAMP. Presynaptic beta$_3$ receptors play a role in the modulation of catecholamine release (Nasser et al. 2006).

Implication of Noradrenaline in CNS Disorders

In Parkinson's and Alzheimer's disease, the locus coeruleus and the medullary noradrenergic cell groups are consistently and profoundly affected (see review by Marien et al. 2004). Cellular loss within the locus coeruleus appears to be more severe in syndromes of dementia. In depression, the implication of the noradrenergic system is based on indirect evidence: reserpine, an inhibitor of vesicular storage of monoamines, can produce depressive symptoms. The majority of clinically effective antidepressant drugs increase the half-life of synaptic noradrenaline by either blocking its reuptake (e.g., imipramine, desipramine, nisoxetine, milnacipran) or blocking its enzymatic degradation (e.g., tranylcypromine).

Receptors	Agonists	Antagonists
β_1	Dobutamine Xamoterol	Atenolol Metoprolol Propranolol
β_2	Salbutamol Terbutaline Salmeterol Zinterol Clenbuterol	Butoxamine Propranolol
α_1	Phenylephrine Oxymetazoline Methoxamine	Tamsulosin Phenoxybenzamine *(irreversible)* Prazosin and analogues: - Doxazocin - Terazoxin - Trimazosin
α_2	Clenbuterol Dexmedetomidine Mivazerol Clonidine Clenbuterol UK 14,304	Yohimbine Idazoxan Atipamezole Efaroxan Dexefaroxan Rauwolscine Phenoxybenzamine *(irreversible)*

Fig. 3.12 Agonists and antagonists of noradrenergic receptors (Kogan et al. 2002; Van Veldhuizen et al. 1993; MacDonald and Scheinin 1995; Ramos and Arnsten 2007; Chen et Minneman 2005; Hague et al. 2003; Saunders and Limbird 1999; Meitzen et al. 2011)

Experimental Tools Influencing the Noradrenergic System

A number of different neurotoxins exist for the noradrenergic system, for example, *N*-(2-chloroethyl)-*N*-ethyl-2-bromobenzylamine (DSP-4) and 6-hydroxydopamine (6-OHDA). However, to improve the selectivity of these neurotoxins, the dopaminergic and serotonergic systems should be protected, respectively, by the concomitant use of specific dopamine and serotonin uptake blockers, such as GBR 12909 and citalopram, respectively. There exist a number of pharmacological tools that are active at different levels of the noradrenergic system, such as amphetamine that acts as a releasing agent and cocaine that blocks noradrenaline reuptake. Caveats for experimental studies using noradrenergic tools are reviewed elsewhere (Marien et al. 2004).

3.2.1.2 Dopamine

In the past, dopamine (Fig. 3.13) was long considered as nothing more than a metabolic intermediate in the synthesis of noradrenaline. Today it is one of the most important neurotransmitters in terms of the diversity and number of functions that it controls and the pathologies in which it is implicated.

Brain Localization

Histochemical techniques have been used to localize dopamine neurons by the revelation of dopamine and its enzyme of synthesis, tyrosine hydroxylase. The central dopaminergic systems are comprised of several nuclei that constitute anatomically distinct systems (Fig. 3.14):

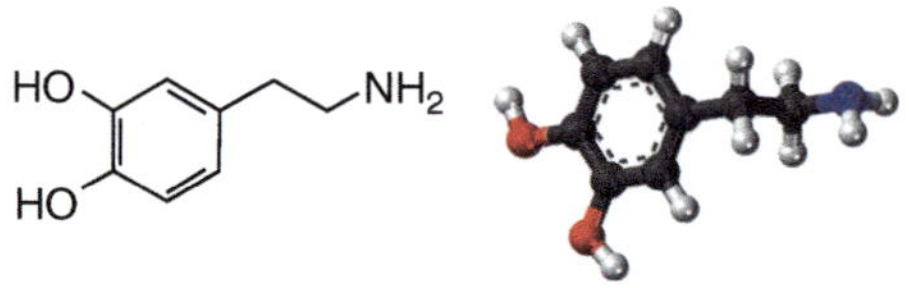

Fig. 3.13 Chemical formula and structure of dopamine

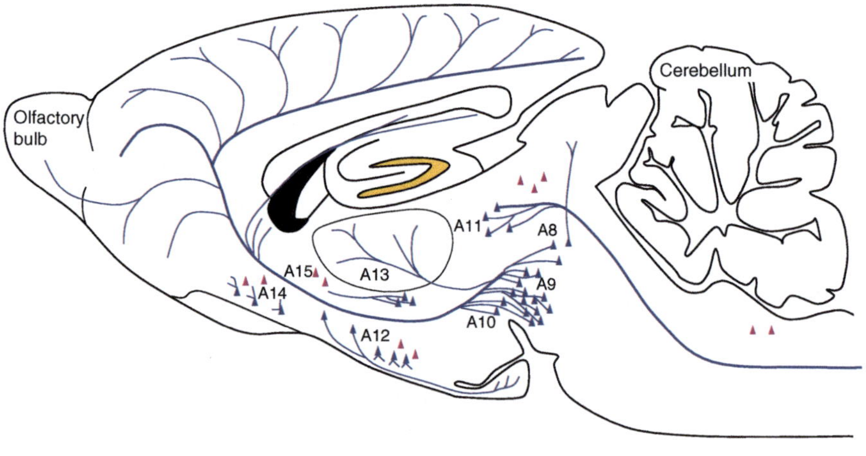

Fig. 3.14 Dopaminergic neurons and their major projections. The *blue triangles* indicate the localization of dopaminergic neurons. *Red triangles* indicate groups of L-dopa-positive cells. *Blue lines* represent dopaminergic fiber projections. A8–A16: dopaminergic cell body groups (Berridge et al. 1997)

- The mesencephalic system (A8, A9 and A10 nuclei)
- The diencephalic system (A11, A12, A13, and A15 nuclei)
- The retinal system (A16 and A17 nuclei, not shown in Fig. 3.14)

The *mesencephalic* system regroups the majority of the dopaminergic cells of the CNS within three juxtaposed structures. The red nucleus, a posterior-lateral extension of the A8 nucleus, and the *pars compacta* (A9) make up the substantia nigra, adjacent to and with some overlap with the ventral tegmental area cell group (A10). The projections from these cell groups are organized into distinct pathways: the nigrostriatal and mesocorticolimbic systems.

The *diencephalic* system is made up of the diencephalo-spinal system (A11), the incerto-hypothalamic system (A13), the periventricular-preoptic system (A14), the tuberoinfundibular and the tuberohypophysial systems (A12), and neurons in the dorsal preoptic area and ventrolateral hypothalamus (A15). Other dopaminergic neuron groups (not shown in Fig. 3.14) include the periglomerular cells of the olfactory bulb (A16) and neurons in the retina (A17).

Biosynthesis and Degradation

The first steps in the synthesis of dopamine from tyrosine are identical to that of noradrenaline. The active uptake of tyrosine by catecholaminergic neurons is followed by its hydroxylation by tyrosine hydroxylase to L-dopa, which in turn is decarboxylated by DOPA decarboxylase to dopamine. Intraneuronal dopamine is concentrated in storage vesicles that are found in large quantity in the presynaptic axon terminals (Fig. 3.15). Several mechanisms contribute to the inactivation of released dopamine: its recapture into the presynaptic terminals via a specific transmembrane transporter, its uptake into glial cells, and its transformation in the synaptic cleft to 3-methoxytyramine (3-MT) by catechol-*O*-methyltransferase (COMT). Within the dopamine neuron, dopamine is metabolized by monoamine oxidase B (MAO-B) to dihydroxyphenylacetic acid (DOPAC). DOPAC that passively diffuses out of the neuron is transformed by COMT to homovanillic acid (HVA).

Receptors

Dopamine receptors are made up of seven transmembrane subunits coupled to a *G* protein. There are two major classes of dopamine receptors (Fig. 3.16): D1 (D1 and D5; D'Aoust and Tiberi 2010) and D2 (D2, D3, and D4; Fukunaga and Shioda 2012; Kvernmo et al. 2008). D1 and D5 receptors are positively coupled to adenylate cyclase by a *Gs* protein, while receptors D2, D3, and D4 are negatively coupled to adenylate cyclase by a *Gi* protein.

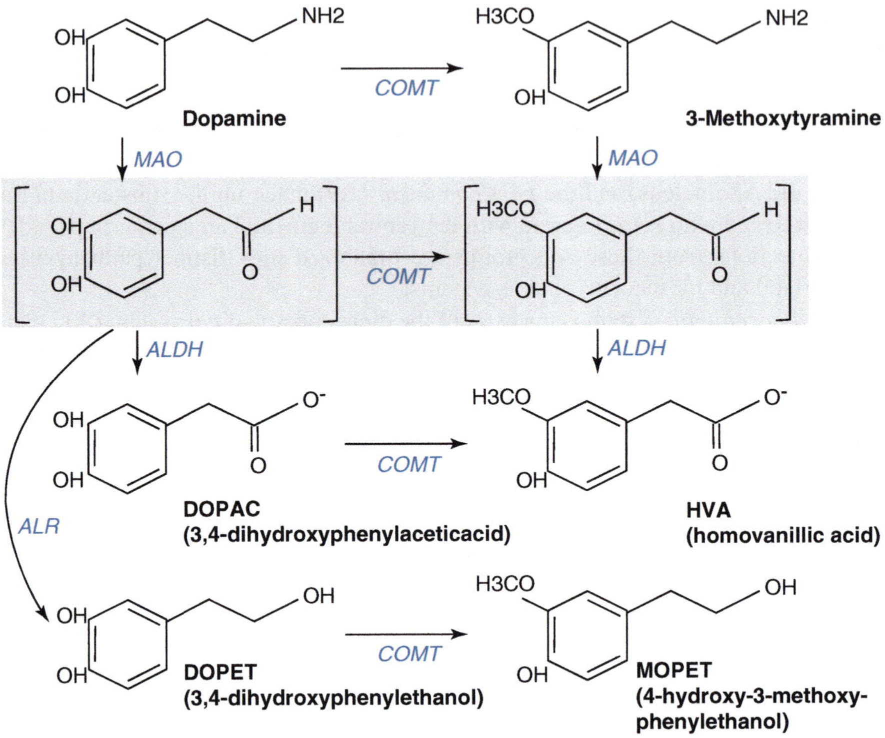

Fig. 3.15 Pathways of dopamine metabolism. The *gray zone* represents an intermediate reaction step. *MAO* monoamine oxidase, *COMT* catechol-*O*-methyltransferase, *ALDH* aldehyde dehydrogenase, *ALR* aldehyde reductase

Implication of Dopamine in CNS Disorders

Given that dopamine participates in a spectrum of functions, it is understandable that modifications in dopaminergic transmission can result in a wide variety of pathologies. Parkinson's disease is an example of dopaminergic hypoactivity that results from the progressive degeneration of nigrostriatal neurons. Conversely, a hyperactive dopaminergic state is associated with the positive symptoms of schizophrenia, such as agitation, delirium, and hallucinations (El Ghundi et al. 2007; Wilson et al. 1998; Joyce and Meador-Woodruff 1997). The mesocorticolimbic pathway plays an important role in reward processes and can constitute a mechanistic base in the addiction to drugs of abuse (Sun 2011; Tuesta et al. 2011; De Biasi and Dani 2011; Sulzer 2011; Schmitt and Reith 2010).

Réceptors		Agonists	Antagonists
D1-like	D1 D5	SKF 38393 SKF 81297 SKF 83959	SCH23390
D2-like	D2	U91,356A Quinpirole Bromocriptine	L–741,626 Raclopride Domperidone Spiperone DR 4004 Eticlopride
	D3	PD 128,907	Nafadotride
	D4	PD 168,077	L–745,870

Fig. 3.16 Synthetic agonists and antagonists of dopamine receptors (Kogan et al. 2002; O'Keefe et al. 2009; Sotoyama et al. 2011; Hoestetler et al. 2011; Huang et al. 2010; Momiyama 2010; Chu et al. 2010; Martelle and Nader 2008)

Experimental Tools Influencing the Dopaminergic System

Different selective toxins for dopaminergic neurons exist, including 1-methyl-4-phenyl-1,2,3,6-tetrahydropyridine (MPTP) which, because of its effect to damage the nigrostriatal system, is often used to model aspects of Parkinson's disease. Another commonly used toxin is 6-hydroxydopamine (6-OHDA) (Zigmond and Stricker, 1984), but its utilization as a selective dopaminergic agent requires protecting against its uptake into and destruction of noradrenergic neurons by pretreatment with desipramine, a blocker of the noradrenergic transporter. As is the case for other transmitter systems, a variety of pharmacological agents exist that are capable of acting on the synthesis, degradation, release, reuptake, and receptor occupancy of dopamine.

3.2.2 Serotonin

Serotonin or 5-hydroxytryptamine (5-HT) (Fig. 3.17) is implicated in the central control of numerous autonomic functions (Martin and Lemonier 1994; Cuche 1981; Berger et al. 2009), motor, sexual and feeding behaviors, pain, and arousal and in affective disorders such as depression and anxiety.

3.2.2.1 Brain Localization

Histochemical techniques have defined the cartography of the serotonergic system by the revelation of serotonin itself, its precursor 5-hydroxytryptophan, and its enzyme of synthesis, tryptophan hydroxylase. The majority of serotonergic nerve cell bodies are situated in nine cell groups within the raphe, arranged in a chain-like fashion in the sagittal plane of the brain stem. These nuclei are numbered from 1 to 9 in the posterior to anterior sense, according to the nomenclature of Dahlstrom and

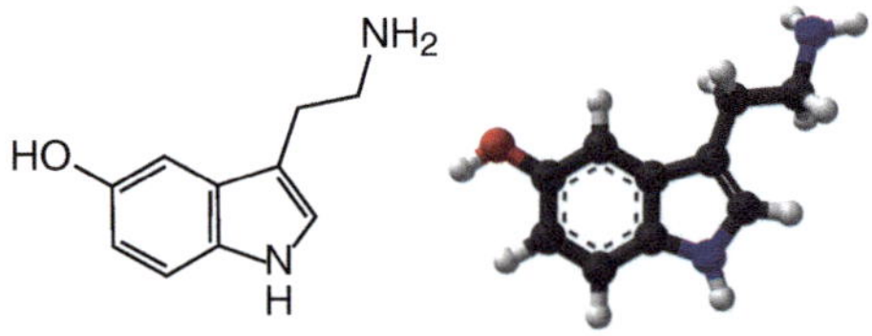

Fig. 3.17 Chemical formula and structure of serotonin

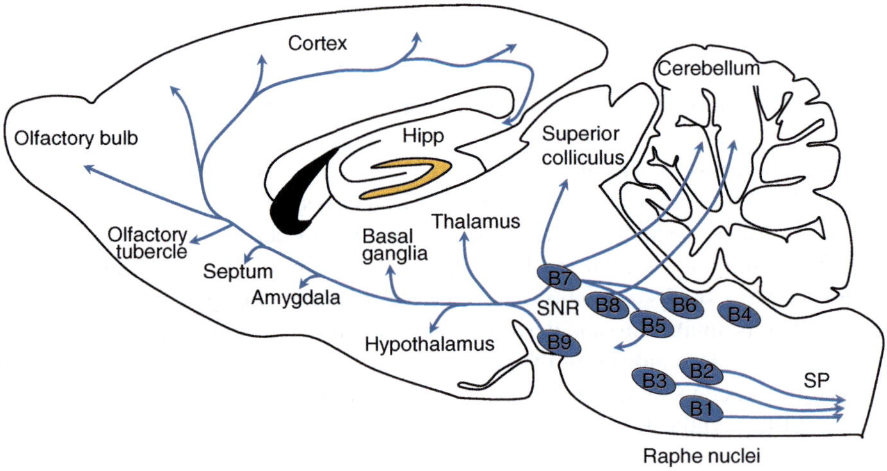

Fig. 3.18 Serotonergic pathways in the CNS. *Hipp* hippocampus, *SNR* substantia nigra reticularis, *SP* spinal cord

Projections	Designation	Structures
Descending projections	B1	Nucleus raphe pallidus Caudal part of the ventrolateral medulla
	B2	Nucleus raphe obscurus
	B3	Nucleus raphe magnus Rostral part of the ventrolateral medulla Lateral paragigantocellular reticular nuclei
	B4	Periventricular grey matter *area postrema*
Ascending projections	B5	Median nucleus raphe pontis
	B6	Dorsal nucleus raphe pontis
	B7	Dorsal raphe nucleus
	B8	Median raphe nucleus Nuceli raphe linearis
	B9	Medial lemniscus

Fig. 3.19 Cell group designations of the ascending and descending projections of the central serotonergic system

Fuxe (1964) (Fig. 3.19). These nuclei also contain non-serotonergic cells. Outside of these nuclei, serotonergic cells are also found in structures such as the dorsomedial nucleus of the hypothalamus.

Serotonergic innervation is extremely diffuse and as a result serotonergic terminals are found in almost all regions of the CNS. Descending fiber projections originate from the B1 and B4 nuclei; ascending pathways originate from the B5 and B9 nuclei (Figs. 3.18 and 3.19).

3.2.2.2 Biosynthesis and Degradation

The precursor of serotonin is tryptophan, an essential amino acid obtained exclusively from dietary sources. The synthesis of serotonin begins by the hydroxylation of tryptophan to 5-hydroxytryptophan by tryptophan hydroxylase. Two cofactors participate in this reaction: oxygen and tetrahydrobiopterin, which act as proton donors. 5-hydroxytryptophan is then decarboxylated by aromatic amino acid

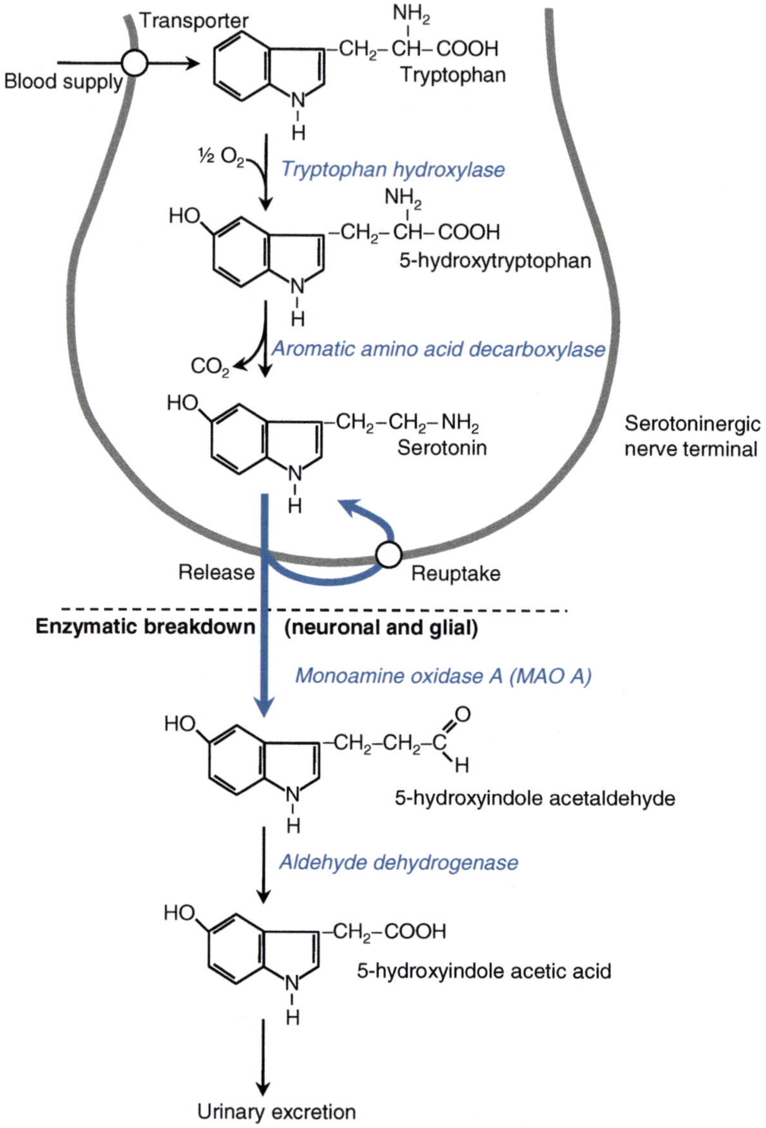

Fig. 3.20 Pathways of synthesis and metabolism of serotonin (From Meunier and Shvaloff 1995)

decarboxylase in the presence of its cofactor vitamin B6 (Fig. 3.20). Serotonin released into the synaptic cleft is inactivated by reuptake via a specific transporter. Selective inhibitors of serotonin reuptake exist, notably fluoxetine (Prozac). Once taken up into the neuron, serotonin is metabolized by monoamine oxidase A (MAO-A), a mitochondrial enzyme. The resulting product, 5-hydroxy-indole-acetaldehyde, is then converted to 5-hydroxy-indoleacetic acid (5-HIAA) by aldehyde dehydrogenase.

3.2.2.3 Receptors

At present, 14 subtypes of serotonergic receptors are known and are grouped into seven classes designated as 5-HT1 to 5-HT7 (Barnes and Sharp 1999) (Fig. 3.21). All classes, with the exception of the 5HT3 receptors, are made up of seven transmembrane units coupled to *G* proteins. There are five subtypes of 5-HT1 receptors – 5-HT1a, 1b, 1d, 1e, and 1f – all of which interact preferentially with *Gi* proteins to inhibit adenylate cyclase activity (Meneses 1999; Cowen 1991). The 5-HT2 class is made up of three subtypes, 5-HT2a, 2b, and 2c, which are all positively coupled to phospholipase C. In contrast to the other serotonergic receptors, the 5-HT3 class belongs to the family of ion channel receptors. It is permeable to sodium and potassium ions, and its stimulation leads to influx of sodium and depolarization (Costall and Naylor 1991).

The 5-HT4 receptor is positively coupled to adenylate cyclase. cAMP produced by the stimulation of this receptor leads to the phosphorylation of diverse proteins and the closure of voltage-dependent potassium channels (Bockaert et al. 2011; King et al. 2008). The 5-HT5 class is made up of two receptor subtypes, 5-HT5a and 5-HT5b, which appear to be coupled to a *G* protein, although the transduction mechanism remains unclear (Nelson 2004). Receptors 5-HT6 and 5-HT7 are both coupled positively to adenylate cyclase (Mahe et al. 2004; Thomas and Hagan 2004; Dawson 2011; Riccioni 2011).

3.2.2.4 Implication in CNS Disorders

A hypoactivity of central serotonergic systems has been highly implicated as a cause of depression, whereas generalized anxiety is associated with an exacerbation of serotonergic transmission. These hypotheses are based largely on the putative mechanisms of action of clinically effective antidepressant and anxiolytic drugs and their effects to modify synaptic levels of serotonin. Partial but significant losses of serotonergic neurons occur in Parkinson's and Alzheimer's disease.

3.2.2.5 Experimental Tools Influencing the Serotonergic System

Serotonergic neurons can be lesioned in a selective manner by 5,7-dihydroxytryptamine (5,7-DHT). However, this requires a protection of noradrenergic and dopaminergic neurons against their uptake of the toxin, by pretreatments with the respective transport inhibitors desipramine and GBR 12909. It is also possible to deplete the serotonergic system using systemic administration of para-chlorophenylalanine (pCPA) or para-chloroamphetamine (pCA). These agents are not entirely selective, however: pCPA can also affect the dopaminergic, noradrenergic, and adrenergic systems. Pharmacological agents that have actions at different levels of the serotonergic system include *d*-fenfluramine (serotonin releaser), fluoxetine (reuptake blocker), and toloxatone (MAO inhibitor).

Receptors		Agonists	Antagonists	
5-HT1	5-HT1a	8-OH–DPAT	WAY 100635 SB 649915 BAY 3702	
	5-HT1b	CP 93129 GR486611X	GR 55562 SB 649915	
	5-HT1d	GR 46611; CP93129	BRL 15572	
	5-HT1e			
	5-HT1f	Ly-334,370		
5-HT2	5-HT2a	DOI; CP93129	MDL 100,907	RP62203 SR46349B SDZSER082
	5-HT2b	CP93129	SB 204741	
	5-HT2c	m-CPP Lorcaserin	RS 102221	
5-HT3 a,b		SR 57227	Ondansetron	
5-HT4 a,b,c,d		BIMU 8 SL65.0155 RS67333 VRX03011 SDZ-216454	GR 113808 RS39604	
5-HT5 (*5-HT1 like*)	5-HT5a		SB-699551-A	
	5-HT5b			
5-HT6		WAY -181187 WAY -208466	Ro 04 - 6790 Dimebolin SB 399885 SB 271046	
5-HT7 a,b,c,d			SB 258719 SB-258741 SB-269970SB → 656104-A DR 4004 → DR-4365;DR-4485	

Fig. 3.21 Synthetic agonists and antagonists of serotonergic receptors (From Borne 1994; Hoyer et al. 1994; Slassi et al. 2004; Forbes et al. 1998; Lovell et al. 2000; Thomas et al. 2003, 2008; Schaffhauser et al. 2009; Hirano et al. 2009; Schechter et al. 2008; King et al. 2008; Restivo et al. 2008; Lucas et al. 2007; Mohler et al. 2007; Mlinar et al. 2006; Parga et al. 2007; Meneses 1999; Thomas 2006; Erlander et al. 1993)

3.3 Acetylcholine

Acetylcholine (Fig. 3.22) was the first substance to which a transmitter role could be attributed. It was shown that acetylcholine was implicated in this role not only at the level of the central nervous system but also at different levels in the peripheral nervous system, in particular at the neuromuscular junction, in smooth muscle and heart, and within ganglionic synapses in the autonomic nervous system.

3.3.1 Brain Localization

The cartography of cholinergic systems have been revealed using antibodies against choline acetyltransferase (ChAT), the enzyme of synthesis of acetylcholine. Acetylcholinesterase (AChE), the enzyme responsible for breakdown of acetylcholine, is also used as a marker of cholinergic neurons, but it is less specific than ChAT since it is also present in non-cholinergic neurons, such as GABAergic interneurons in the striatum that also co-express somatostatin. The central cholinergic systems are classified into two general categories: isolated or small groups of neurons and larger groups of cells that form cholinergic nuclei. The first category includes the cholinergic motoneurons of the spinal cord, preganglionic and postganglionic parasympathetic cells of the autonomic nervous system, and interneurons best exemplified by the cholinergic interneurons of the striatum. The second category is made up of eight groups of cholinergic neurons, divided into groups designated Ch1 to Ch8 according to the classification by Mesulam (1998) (Fig. 3.23):

- Groups Ch1 and Ch2 correspond respectively to cholinergic neurons of the medial septal nucleus and the vertical limb of the diagonal band of Broca. These two nuclei make up the principal source of the septo-hippocampal pathway.
- Group Ch3 is situated in the horizontal limb of the diagonal band of Broca and innervates the olfactory bulb and cingular cortex.
- Group Ch4 is made up of the magnocellular preoptic nucleus, the substantia innominata, and the nucleus basalis magnocellularis (basal nucleus of Meynert in the human brain). These neurons innervate principally the cerebral cortex and amygdala.
- Groups Ch5 and Ch6 are located in the brain stem and correspond, respectively, to the pedunculopontine and latero-dorsal tegmental nuclei. These neurons are the principal source of ascending cholinergic projections to thalamic and hypothalamic structures. They also provide descending projections to pontine, deep cerebellar, spinal, and vestibular nuclei, the locus coeruleus, and the raphe magnus. A minor portion of the pedunculopontine cell group also projects to the cerebral cortex.

Fig. 3.22 Chemical formula and structure of acetylcholine

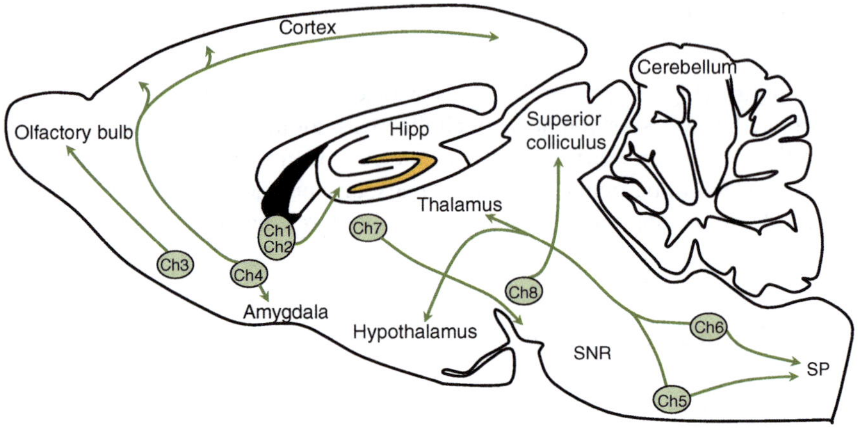

Fig. 3.23 Cholinergic pathways in the CNS. *Hipp* hippocampus, *SNR* substantia nigra reticularis, *SP* spinal cord

- Group Ch7, located in the medial habenular nucleus, innervates the interpeduncular nucleus.
- Group Ch8, comprising the parabigeminal nucleus, innervates in a preferential manner the superior colliculus.

3.3.2 Biosynthesis and Degradation (Fig. 3.24)

Acetylcholine is synthesized by choline acetyltransferase from acetyl-coenzyme-A (acetyl-CoA) and choline. Acetyl-CoA is produced in mitochondria from pyruvate generated by the Krebs cycle. Brain choline is derived from circulating sources in the plasma, from the hydrolysis of phospholipids, or from the enzymatic breakdown of acetylcholine released into the synapse. Acetylcholine is stored in vesicular and non-vesicular compartments. Vesamicol is an agent capable of blocking the specific transporter for acetylcholine on vesicular membranes. Released acetylcholine is eliminated from the synaptic cleft by its breakdown by acetylcholinesterase (AChE) located on pre- and postsynaptic membranes. Choline thus formed is taken up by the neuron by transport systems that have both high and low affinity for hemicholinium-3 and that are found exclusively on cholinergic nerve terminals. Several inhibitors of AChE exist, most notably tacrine, physostigmine, galanthamine (which also has some nicotinic receptor agonist properties), neostigmine, and sarin nerve gas.

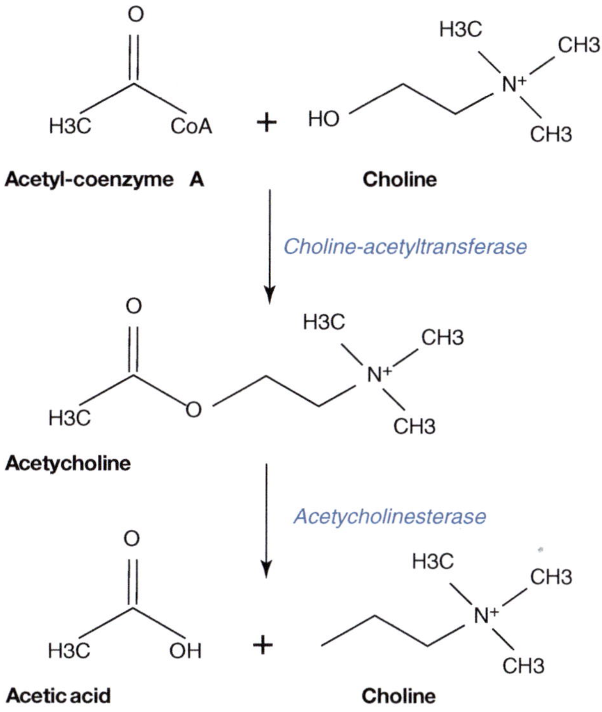

Fig. 3.24 Pathways of biosynthesis and degradation of acetylcholine

3.3.3 Receptors

Released acetylcholine binds to two types of receptors (Fig. 3.25), ionotropic receptors referred to as nicotinic since they are sensitive to nicotine and metabotropic receptors referred to as muscarinic since they are sensitive to muscarine.

3.3.3.1 Nicotinic Receptors

Nicotinic receptors are ion channel receptors made up of five subunits organized in a rosette pattern around the ion channel (Millar and Giotti 2009). Only two types of subunits exist, alpha and beta. The alpha subunit itself has nine subtypes (alpha2 to alpha10) and the beta subunit 3 subtypes (beta2 to beta4). In 2009, 17 subunits were known and sequenced: alpha1 to alpha10, beta1 to beta4, gamma, delta, and epsilon. Of the considerable number of pentamers possible, only a dozen

Receptors	Agonists	Antagonists
Nicotinic	Nicotine	Hexamethonium
Neuronal	Anatoxin-a	Mecamylamine
a_7	DMXB-A Tropisetron	Methyllycaconitine
a_4 b_2	Varenicline	DMXB-A
a_6		Alpha-conotoxin MII
Muscarinic	Muscarine Pilocarpine	Scopolamine Oxitropium Atropine
M1	Pilocarpine	Pirenzepine
M2	Bethanechol	Methoctramine
M3	L–689,660	4-DAMP hexahydro-sila-diffénidol (HHSiD)
M4	McN-A-343	Himbacine Tropicamide
M5		

Fig. 3.25 Agonists and antagonists of cholinergic receptors (Ishikawa and Hashimoto 2011; Crunelle et al. 2010; Gotti and Clementi 2004; Kem 2000)

have actually been characterized. Molecular biological studies have revealed several types of central nicotinic receptors that are functionally and pharmacologically distinct depending on the composition and arrangement of their subunits (Ishikawa and Hashimoto 2011; Haydar and Dunlop 2010; Gotti and Clementi 2004; Gotti et al. 2009). Pentamers making up the nicotinic receptors have at least two alpha subunits, each equipped with a receptor-binding site for acetylcholine. The binding of two acetylcholine molecules on two receptor sites on the alpha subunits induces a conformational change of the pentamer and the opening of the ion channel that allows the cellular influx of sodium and calcium ions and consequently membrane depolarization. The ion channel is also permeable to potassium ions. The presence of the beta2 subunit is essential for the functioning of the receptor and appears to play a major role in the stimulant effects of nicotine.

3.3.3.2 Muscarinic Receptors

Muscarinic receptors are members of the receptor family composed of seven transmembrane domains coupled to a *G* protein. They exist as five subtypes, M1 to M5 (Reever et al. 1997). M1, M3, and M5 receptors are positively coupled to phospholipase C via a *Gs* or *Gq* protein. Their activation leads to the hydrolysis of membrane phosphatidylinositol diphosphate to inositol triphosphate and increases the cytosolic concentration of calcium and diacylglycerol to activate protein kinase C. M2 and M4 receptors are negatively coupled to adenylate cyclase via a *Gi* or *Go* protein. Their activation leads to a decrease in cAMP levels and consequently the inactivation of cAMP-dependent proteins. The activation of M1 and M3 receptors may in certain cases also induce the closure of M-type potassium channels, causing depolarization. The activation of M2 and M4 receptors may also modulate the opening of certain potassium channels.

3.3.3.3 Implication in CNS Disorders

Central cholinergic dysfunction occurs over the course of aging in patients suffering from dementia associated with neurodegenerative disorders such as Alzheimer's disease (Parri et al. 2011; Wallace and Porter 2011; Kawamata and Shimohama 2011; Schliebs and Arendt 2011; Quik et al. 2009; Langmead et al. 2008; Palma et al. 2012). Cholinergic structures and in particular region Ch4 show several types of lesions characterized by accumulation of amyloid protein, senile plaques, and neurofibrillary tangles. Parkinson's disease is characterized by a hyperactivity of striatal cholinergic interneurons in response to a deficit in nigrostriatal dopaminergic afferent control. Before the advent of L-dopa, cholinergic antagonists such as trihexyphenidyl were the first pharmacotherapies used to treat motor symptoms of the disease. Huntington's disease is characterized by a progressive loss of striatal cholinergic interneurons.

3.3.3.4 Experimental Tools Influencing the Cholinergic System

192-IgG-saporin is an immunotoxin composed of an antibody and a protein that is selective for cholinergic cells. The 192-IgG antibody specifically recognizes the low-affinity nerve growth factor receptor p75NTR present on cholinergic neurons in cell groups Ch1 and Ch4, except for those that project to the amygdala. Saporin is a protein that disrupts ribosome function. When 192-IgG binds to the p75NTR receptor, the toxin-receptor complex is internalized by endocytosis and transported in a retrograde fashion towards the cell body. There, the saporin component dissociates from the antibody inside the endosome, ribosomes are inactivated, protein synthesis is consequently compromised, and cell death ensues.

References

Barnes NM, Sharp T (1999) A review of central 5-HT receptors and their function. Neuropharmacology 38(8):1083–1152

Berger P, Farrel K, Sharp F, Skolnick P (1994) Drugs acting at the strychnine-insensitive glycine receptor do not induce HSP-70 protein in the cingulate cortex. Neurosci Lett 168(1–2): 147–150

Berger M, Gray JA, Roth BL (2009) The expanded biology of serotonin. Annu Rev Med 60:355–366

Berridge CW, Stratford TL, Foote SL, Kelley AE (1997) Distribution of dopamine-beta-hydroxylase-like immunoreactive fibers within the shell subregion of the nucleus accumbens. Synapse 27:230–241

Binns KE, Turner JP, Salt TE (2003) Kainate receptor (GluR5) – mediated disinhibition of responses in rat ventrobasal thalamus allows a novel sensory processing mechanism. J Physiol 551(Pt 2):525–537

Bjorklund A, Hokfelt T, Kuhar MJ (2005) The handbook of chemical neuroanatomy. Elsevier, Amsterdam

Bockaert J, Claeysen S, Compan V, Dumuis A (2011) 5-HT(4) receptors, a place in the sun: act two. Curr Opin Pharmacol 11(1):87–93

Bolea S, Avignone E, Berretta N, Sanchez-Andres JV, Cherubini E (1999) Glutamate controls the induction of GABA-mediated giant depolarizing potentials through AMPA receptors in neonatal rat hippocampal slices. J Neurophysiol 81(5):2095–2102

Bon C, Galvan M (1996) Electrophysiological actions of GABAB agonists and antagonists in rat dorso-lateral septal neurones in vitro. Br J Pharmacol 118(4):961–967

Borne RF (1994) Serotonin: the neurotransmitter for the'90s. Drug Topics 10:108–120

Chebib M, Hinton T, Schmid KL et al (2009) Novel, potent, and selective $GABA_C$ antagonists inhibit myopia development and facilitate learning and memory. J Pharmacol Exp Ther 328(2): 448–457

Chen ZJ, Minneman KP (2005) Recent progress in $alpha_1$-adrenergic receptor research. Acta Pharmacol Sin 26(11):1281–1287

Chu HY, Yang Z, Zhao B, Jin GZ, Hu GY, Zhen X (2010) Activation of phosphatidyl-inositol-linked D1-like receptors increases spontaneous glutamate release in rat somatosensory cortical neurons in vitro. Brain Res 1343:20–27

Cooper JR, Bloom FL, Roth RH (2003) The biochemical basis of neuropharmacology, 8th edn. Oxford Academy Press, Oxford

Costall B, Naylor RJ (1991) Pharmacological properties and functions of central 5-HT3 receptors. Therapie 46(6):437–444

Cowen PJ (1991) Serotonin receptor subtypes: implications for psychopharmacology. Br J Psychiat 12:7–14

Crunelle CL, Miller ML, Booij J, van den Brink W (2010) The nicotinic acetylcholine receptor partial agonist varenicline and the treatment of drug dependence: a review. Eur Neuropsychopharmacol 20(2):69–79

Cuche H (1981) *Séminaire de psychiatrie biologique, Hôpital Sainte-Anne*, tome 1. Edition Medicales Fournier Freres, Gennevilliers, pp 107–125

D'Aoust JP, Tiberi M (2010) Role of the extracellular amino terminus and first membrane spanning helix of dopamine D1 and D5 receptors in shaping ligand selectivity and efficacy. Cell Signal 22(1):106–116

Dahlstrom A, Fuxe K (1964) Localization of monoamines in the lower brain stem. Experientia 20:398–399

Danysz W, Parsons CG (1998) Glycine and *N*-methyl-D-aspartate receptors: physiological significance and possible therapeutic applications. Pharmacol Rev 50(4):597–664

Dawson LA (2011) The central role of 5-HT6 receptors in modulating brain neurochemistry. Int Rev Neurobiol 96:1–26

De Biasi M, Dani JA (2011) Reward, addiction, withdrawal to nicotine. Annu Rev Neurosci 34:105–130
El-Ghundi M, O'Dowd BF, George SR (2007) Insights into the role of dopamine receptor systems in learning and memory. Rev Neurosci 18(1):37–66
Epelbaum J (1995) Neuropeptides et neuromédiateurs, 2eth edn. Editions Inserm/Sandoz, Paris
Erlander MG, Lovenberg TW, Baron BM et al (1993) Two members of a distinct subfamily of 5-hydroxytryptamine receptors differentially expressed in rat brain. Proc Natl Acad Sci U S A 90(8):3452–3456
Forbes IT, Dabbs S, Duckworth DM et al (1998) (R)-3,*N*-dimethyl158 *N*-[1-methyl-3-(4-methyl-piperidinyl-1-yl) propyl]benzensulfonamide: the first selective 5-HT7 antagonist. J Med Chem 41:655–657
Fukunaga K, Shioda N (2012) Novel dopamine D2 receptor signaling through proteins interacting with the third cytoplasmic loop. Mol Neurobiol 45(1):144–152
Funahashi M, Stewart M (1998) Properties of gamma-frequency oscillations initiated by propagating population bursts in retrohippocampal regions of rat brain slices. J Physiol 510(Pt 1):191–208
Gether U, Asmar F, Meinild AK, Rasmussen SG (2002) Structural basis for activation of G-protein-coupled receptors. Pharmacol Toxicol 91(6):304–312
Gotti C, Clementi F (2004) Neuronal nicotinic receptors: from structure to pathology. Prog Neurobiol 74:363–396
Gotti C, Clementi F, Fornari A et al (2009) Structural and functional diversity of native brain neuronal nicotinic receptors. Biochem Pharmacol 78(7):703–711
Hague C, Chen Z, Uberti M, Minneman KP (2003) Alpha(1)-adrenergic receptor subtypes: non-identical triplets with different dancing partners ? Life Sci 74(4):411–418
Haydar SN, Dunlop J (2010) Neuronal nicotinic acetylcholine receptors – targets for the development of drugs to treat cognitive impairment associated with schizophrenia and Alzheimer's disease. Curr Top Med Chem 10(2):144–152
Hirano K, Piers TM, Searle KL, Miller ND, Rutter AR, Chapman PF (2009) Procognitive 5-HT6 antagonists in the rat forced swimming test: potential therapeutic utility in mood disorders associated with Alzheimer's disease. Life Sci 84(15–16):558–562
Hostetler CM, Harkey SL, Krzywosinski TB, Aragona BJ, Bales KL (2011) Neonatal exposure to the D1 agonist SKF38393 inhibits pair bonding in the adult prairie vole. Behav Pharmacol 22(7):703–710
Hoyer D, Clarke DE, Fozard JR et al (1994) International union of pharmacology: classification of receptors for 5 hydroxytryptamine (serotonin). Pharmacol Rev 46:157–205
Huang Y, Qiu AW, Peng YP, Liu Y, Huang HW, Qiu YH (2010) Roles of dopamine receptor subtypes in mediating modulation of T lymphocyte function. Neuro Endocrinol Lett 31(6):782–791
Ishikawa M, Hashimoto K (2011) Alpha7 nicotinic acetylcholine receptor as a potential therapeutic target for schizophrenia. Curr Pharmacol Des 17(2):121–129
Joyce JN, Meador-Woodruff JH (1997) Linking the family of D2 receptors to neuronal circuits in human brain: insights into schizophrenia. Neuropsychopharmacology 16(6):375–384
Kabashima N, Shibuya I, Ibrahim N, Ueta Y, Yamashita H (1997) Inhibition of spontaneous EPSCs and IPSCs by presynaptic $GABA_B$ receptors on rat supraoptic magnocellular neurons. J Physiol 504(Pt 1):113–126
Kawamata J, Shimohama SS (2011) Stimulating nicotinic receptors trigger multiple pathways attenuating cytotoxicity in models of Alzheimer's and Parkinson's diseases. J Alzheimers Dis 24(Suppl 2):95–109
Kem WR (2000) The brain alpha7 nicotinic receptor may be an important therapeutic target for the treatment of Alzheimer's disease: studies with DMXBA (GTS-21). Behav Brain Res 113(1–2): 169–181
King MV, Marsden CA, Fone KC (2008) A role for the 5-HT(1A), 5-HT4 and 5-HT6 receptors in learning and memory. Trends Pharmacol Sci 29(9):482–492
Kogan HA, Marsden CA, Fone KC (2002) DR4004, a putative 5-HT(7) receptor antagonist, also has functional activity at the dopamine D2 receptor. Eur J Pharmacol 449(1–2):105–111

Kvernmo T, Houben J, Sylte I (2008) Receptor-binding and pharmacokinetic properties of dopaminergic agonists. Curr Top Med Chem 8(12):1049–1067
Labrakakis C, Tong CK, Weissman T, Torsney C, MacDermott AB (2003) Localization and function of ATP and $GABA_A$ receptors expressed by nociceptors and other postnatal sensory neurons in rat. J Physiol 549(Pt 1):131–142
Langmead CJ, Watson J, Reavill C (2008) Muscarinic acetylcholine receptors as CNS drug targets. Pharmacol Ther 117(2):232–243
Lovell PJ, Bromidge SM, Dabbs S et al (2000) A novel, potent, and selective 5-HT7 antagonist: (R)-3-(2-(2-(4-methylpiperin-1-yl)-ethyl)pyrrolidine-1-sulfonyl)phenol (SB-269970). J Med Chem 43:342–345
Lucas G, Rymar VV, Du J et al (2007) Serotonin(4) (5-HT(4)) receptor agonists are putative antidepressants with a rapid onset of action. Neuron 55(5):712–725
MacDonald E, Scheinin M (1995) Distribution and pharmacology of alpha 2-adrenoceptors in the central nervous system. J Physiol Pharmacol 46(3):241–258
Mahe C, Loetscher E, Feuerbach D, Muller W, Seiler MP, Schoeffter P (2004) Differential inverse agonist efficacies of SB-258719, SB-258741 and SB-269970 at human recombinant serotonin 5-HT7 receptors. Eur J Pharmacol 495(2–3):97–102
Marien MR, Colpaert FC, Rosenquist AC (2004) Noradrenergic mechanisms in neurodegenerative diseases: a theory. Brain Res Rev 45:38–78
Martelle JL, Nader MA (2008) A review of the discovery, pharmacological characterization, and behavioral effects of the dopamine D2-like receptor antagonist eticlopride. CNS Neurosci Ther 14(3):248–262
Martin P, Lemonier F (1994) Interet des recepteurs serotoninergiques de type 2: 5HT 2a et 5HT 2c, dans les troubles depressifs: action de la medifoxamine. L'encéphale 20:427–435
Meitzen J, Luoma JI, Stern CM, Mermelstein PG (2011) Alpha1-adrenergic receptors activate two distinct signaling pathways in striatal neurons. J Neurochem 116(6):984–995
Meneses A (1999) 5-HT system and cognition. Neurosci Biobehav Rev 23(8):1111–1125
Mesulam MM (1998) From sensation to cognition. Brain 121(6):1013–1052
Meunier JM, Shvaloff A (1995) Neurotransmetteurs: bases neurobiologiques et pharmacologiques. Editions Masson, Paris
Millar NS, Gotti C (2009) Diversity of vertebrate nicotinic acetylcholine receptors. Neuropharmacology 56(1):237–246
Mirza NR, Larsen JS, Mathiasen C et al (2008) NS11394 [3'-[5-(1-hydroxy-1-methyl-ethyl)-benzoimidazol-1-yl]-biphenyl-2-carbonitrile], a unique subtype-selective $GABA_A$ receptor positive allosteric modulator: in vitro actions, pharmacokinetic properties and in vivo anxiolytic efficacy. J Pharmacol Exp Ther 327(3):954–968
Mlinar B, Mascalchi S, Mannaioni G, Morini R, Corradetti R (2006) 5-HT4 receptor activation induces long-lasting EPSP-spike potentiation in CA1 pyramidal neurons. Eur J Neurosci 24(3):719–731
Mohler EG, Shacham S, Noiman S et al (2007) VRX-03011, a novel 5-HT4 agonist, enhances memory and hippocampal acetylcholine efflux. Neuropharmacology 53(4): 563–573
Momiyama T (2010) Developmental increase in D1-like dopamine receptor-mediated inhibition of glutamatergic transmission through P/Q-type channel regulation in the basal forebrain of rats. Eur J Neurosci 32(4):579–590
Nasser Y, Ho W, Sharkey KA (2006) Distribution of adrenergic receptors in the enteric nervous system of the guinea pig, mouse, and rat. J Comp Neurol 495(5):529–553
Nelson DL (2004) 5-HT5 receptors. Current drug targets. CNS Neurol Disord 3(1):53–58
O'Keefe GC, Barker RA, Caldwell MA (2009) Dopaminergic modulation of neurogenesis in the subventricular zone of the adult brain. Cell Cycle 8(18):2888–2894
Palma E, Conti L, Roseti C, Limatola C (2012) Novel approaches to study the involvement of a7-nAChR in human diseases. Curr Drug Targets 13:579–586
Parga J, Rodriguez-Pallares J, Munoz A, Guerra MJ, Labandeira-Garcia JL (2007) Serotonin decreases generation of dopaminergic neurons from mesencephalic precursors via serotonin type 7 and type 4 receptors. Dev Neurobiol 67(1):10–22

Parri HR, Hernandez CM, Dineley KT (2011) Research update: alpha7 nicotinic acetylcholine receptor mechanisms in Alzheimer's disease. Biochem Pharmacol 82(8):931–942

Parsons CG (2001) NMDA receptors as targets for drug action in neuropathic pain. Eur J Pharmacol 429(1–3):71–78

Patenaude C, Chapman CA, Bertrand S, Congar P, Lacaille JC (2003) $GABA_B$ receptor- and metabotropic glutamate receptor-dependent cooperative long-term potentiation of rat hippocampal $GABA_A$ synaptic transmission. J Physiol 553(Pt 1):155–167

Quik M, Huang LZ, Parameswaran N, Bordia T, Campos C, Perez XA (2009) Multiple roles for nicotine in Parkinson's disease. Biochem Pharmacol 78(7):677–685

Ramos BP, Arnsten AF (2007) Adrenergic pharmacology and cognition: focus on the prefrontal cortex. Pharmacol Ther 113(3):523–536

Reever CM, Ferrari-DiLeo G, Flynn DD (1997) The M5 (m5) receptor subtype: fact or fiction ? Life Sci 60(13–14):1105–1112

Reis GM, Duarte ID (2006) Baclofen, an agonist at peripheral GABAB receptors, induces antinociception via activation of TEA-sensitive potassium channels. Br J Pharmacol 149(6): 733–739

Restivo L, Roman F, Dumuis A, Bockaert J, Marchetti E, Ammassari-Teule M (2008) The promnesic effect of G-protein-coupled 5-HT4 receptors activation is mediated by a potentiation of learning induced spine growth in the mouse hippocampus. Neuropsychopharmacology 33(10):2427–2434

Riccioni T (2011) 5-HT6 receptor characterization. Int Rev Neurobiol 94:67–88

Saunders C, Limbird LE (1999) Localization and trafficking of alpha2-adrenergic receptor subtypes in cells and tissues. Pharmacol Ther 84(2):193–205

Schaffhauser H, Mathiasen JR, Dicamillo A et al (2009) Dimebolin is a 5-HT6 antagonist with acute cognition enhancing activities. Biochem Pharmacol 78(8):1035–1042

Schechter LE, Lin Q, Smith DL et al (2008) Neuropharmacological profile of novel and selective 5-HT6 receptor agonists: WAY-181187 and WAY-208466. Neuropsychopharmacology 33(6): 1323–1335

Schliebs R, Arendt T (2011) The cholinergic system in aging and neuronal degeneration. Behav Brain Res 221(2):555–563

Schmitt KC, Reith ME (2010) Regulation of the dopamine transporter: aspects relevant to psychostimulant drugs of abuse. Ann NY Acad Sci 1187:316–340

Shen KZ, Johnson SW (1997) Presynaptic GABAB and adenosine A1 receptors regulate synaptic transmission to rat substantia nigra reticulata neurones. J Physiol 505(1):153–163

Slassi A, Methvin I, Xin T (2004) Recent progress in 5-HT7 receptors: potential treatment of central and peripheral nervous system diseases. Expert Opin Ther Pat 14:1009–1027

Sotoyama H, Zheng Y, Iwakura Y et al (2011) Pallidal hyperdopaminergic innervation underlying D2 receptor-dependent behavioral deficits in the schizophrenia animal model established by EGF. PLoS One 6(10):e25831

Sulzer D (2011) How addictive drugs disrupt presynaptic dopamine neurotransmission. Neuron 69(4):628–649

Sun W (2011) Dopamine neurons in the ventral tegmental area: drug-induced synaptic plasticity and its role in relapse to drug-seeking behavior. Curr Drug Abuse Rev 4(4):270–285

Sun L, Chiu D, Kowal D et al (2004) Characterization of two novel *N*-methyl-D-aspartate antagonists: EAA-090 (2-[8,9-dioxo-2,6-diazabicyclo [5.2.0]non-1(7)-en2-yl]ethylphosphonic acid) and EAB-318 (R-alpha-amino-5-chloro-1-(phosphonomethyl)-1*H*-benzimidazole-2-propanoic acid hydrochloride). J Pharmacol Exp Ther 310(2):563–750

Thomas DR (2006) 5-ht5A receptors as a therapeutic target. Pharmacol Ther 111(3):707–714

Thomas DR, Hagan JJ (2004) 5-HT7 receptors. Current drug targets. CNS Neurol Disord 3:81–90

Thomas DR, Melotto S, Massagrande M et al (2003) SB-656104-A, a novel selective 5-HT7 receptor antagonist, modulates REM sleep in rats. Br J Pharmacol 139:705–714

Thomsen WJ, Grottick AJ, Menzaghi F et al (2008) Lorcaserin, a novel selective human 5-hydroxytryptamine2C agonist: in vitro and in vivo pharmacological characterization. J Pharmacol Exp Ther 325(2):577–587

Tuesta LM, Fowler CD, Kenny PJ (2011) Recent advances in understanding nicotinic receptor signaling mechanisms that regulate drug self-administration behavior. Biochem Pharmacol 82(8):984–995

Van Veldhuizen MJA, Feenstra MGP, Heinsbroek RPW, Boer GJ (1993) In vivo microdialysis of noradrenaline overflow: effects of alpha-adrenoceptor agonists and antagonists measured by cumulative concentration response curves. Br J Pharmacol 109:655–660

Wallace TL, Porter RH (2011) Targeting the nicotinic alpha7 acetylcholine receptor to enhance cognition in disease. Biochem Pharmacol 82(8):891–903

Weiss B, Alt A, Ogden AM et al (2006) Pharmacological characterization of the competitive GLUK5 receptor antagonist decahydroisoquinoline LY466195 in vitro and in vivo. J Pharmacol Exp Ther 318(2):772–781

Wilson JM, Sanyal S, Van Tol HH (1998) Dopamine D2 and D4 receptor ligands: relation to antipsychotic action. Eur J Pharmacol 351(3):273–286

Xie A, Song X, Ripps H, Qian H (2008) Cyclothiazide: a subunit-specific inhibitor of $GABA_C$ receptors. J Physiol 586(Pt 11):2743–2752

Yang L, Omori K, Omori K, Otani H, Suzukawa J, Inagaki C (2003) GABAC receptor agonist suppressed ammonia-induced apoptosis in cultured rat hippocampal neurons by restoring phosphorylated BAD level. J Neurochem 87(3):791–800

Yang JC, Fan XL, Song XA, Li Q (2008) The role of different glutamate receptors in the mediation of glutamate-evoked excitation of red nucleus neurons after simulated microgravity in rat. Neurosci Lett 448(3):255–259

Zigmond MJ, Stricker EM (1984) Parkinson's disease: studies with an animal model. Life Sci 35(1):5–18

Chapter 4
Stereotaxic Approach of a Target Structure

"Stereotaxy" is defined as "a method in neurosurgery and neurological research for locating points within the brain using an external, three-dimensional frame of reference usually based on the Cartesian coordinate system" (*American Heritage Dictionary of the English Language* http://www.ahdictionary.com/word/search.html?q=stereotaxy). Stereotaxic surgery has thus been described as "the introduction of instruments, or the direction of radiation beams into a small, well defined but invisible target within the skull… to explore the functional properties of this target, to biopsy it or to alter it by physical or chemical means" (De La Porte 1993).

The stereotaxic approach implies a knowledge and understanding of the anatomical and functional characteristics of brain structures, the spatial coordinates of the target, and the basic principles of surgery related to anesthesia, analgesia, and asepsis. The quality of the end result of the operation and its reproducibility depend on the respect of a precise and rigorous procedure. The principal notions necessary to undertake this type of surgery are closely associated with regulatory, bibliographic, technical, and mathematical aspects.

4.1 The Stereotaxic Atlas

The implantation of cannulae, electrodes, or other instruments into a given brain structure requires a knowledge of the theoretical coordinates of implantation. It is the responsibility of the user to be particularly vigilant as to the adequacy of the stereotaxic coordinates shown in a published atlas and those that should otherwise be verified and validated for use in his/her own experimental model.

In this regard, the body weight, sex, species, strain, and age of the animals to be operated are critical factors to be taken into account in order to adjust the stereotaxic coordinates proposed by an atlas, which are specified for a particular strain of the species and weight. The investigator should be aware of the peculiarities specific to each atlas and to choose the one best suited to his/her needs.

B. Ferry et al., *Stereotaxic Neurosurgery in Laboratory Rodent*,
DOI 10.1007/978-2-8178-0472-9_4,

Several stereotaxic atlases of the rat and mouse brain are currently available. Since these atlases use different cartographic parameters, the coordinates determined using one atlas are valid only for that particular atlas The examples of coordinates cited in this guide are based on the atlas of Paxinos and Watson (4th edition, 1998 and 5th edition, 2004).

4.1.1 Stereotaxic Atlases

4.1.1.1 Rat

Different atlases listing the brain structures in the rat can be found. Some of them are listed below:

- König JFR, Klippel RA (1963) The rat brain: a stereotaxic atlas of the forebrain and the lower parts of the brain stem
- Albe-Fessard D, Stutinsky F, Libouan S (1966) Atlas stéréotaxique du diencéphale du rat blanc
- Pellegrino LJ, Cushman AJ (1967) A stereotaxic atlas of the rat brain
- Cooley RK (1990) Stereotaxic surgery in the rat: a photographic series
- Altman J, Bayer SA (1994) Atlas of prenatal rat brain development
- Swanson L (2003) Brain maps: structure of the rat brain
- Paxinos G, Watson C (1998) The rat brain in stereotaxic coordinates, 4th edn
- Paxinos G, Watson C (2004) The rat brain in stereotaxic coordinates, 5th edn
- Paxinos G, Watson C (2007) The rat brain in stereotaxic coordinates, 6th edn
- Paxinos G, Watson C (2013) The rat brain in stereotaxic coordinates, 7th edn

4.1.1.2 Mouse

Different atlases listing the brain structures in the mouse can be found. Some of them are listed below:

- Sidman RL, Angevine JB Jr, Taber PE (1971) Atlas of the mouse brain and spinal cord
- Montemurro DG, Dukelow RH (1971) A stereotaxic atlas of the diencephalon and related structures of the mouse
- Lehmann A, Gautier M, Ghilini G, Henry J, Langlois R, Laplante S (1974) Atlas stéréotaxique du cerveau de la souris
- Schambra UB, Lauder JM, Silver J (1991) Atlas of the prenatal mouse brain
- Paxinos G, Franklin KBJ (2001) The mouse brain in stereotaxic coordinates, 2nd edn
- Paxinos G, Franklin KBJ (2003) The mouse brain in stereotaxic coordinates: compact, 2nd edn

- Franklin KBJ, Paxinos G (2008) The mouse brain in stereotaxic coordinates, 3rd edn
- Paxinos G, Franklin KBJ (2012) The mouse brain in stereotaxic coordinates, 4th edn

4.2 The Stereotaxic Apparatus

The stereotaxic apparatus comprises a U-shaped stereotaxic frame mounted on a baseplate. Three mechanical elements (microdrives) fixed upon the frame allow travel of an electrode (or cannula) holder along three orthogonal axes: forward and backward, up and down, and side to side (Fig. 4.1). This micromanipulator assembly

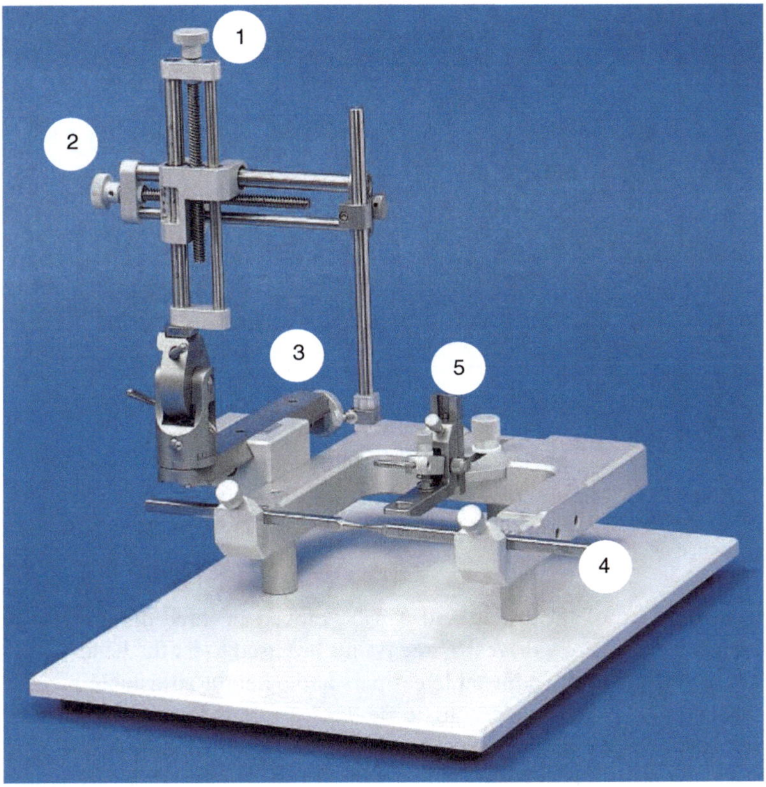

Fig. 4.1 Stereotaxic frame equipped with a micromanipulator. (*1*) Microdrive screw adjustment of the dorso-ventral axis; (*2*) microdrive screw adjustment of the mediolateral axis; (*3*) microdrive screw adjustment of the antero-posterior axis; (*4*) ear bars to center the head parallel to the antero-posterior axis of the frame; (*5*) incisor bar, which determines and fixes the plane of inclination of the head in the "flat-skull position," such that the bregma and lambda points are located on the same horizontal plane, i.e., to have the same dorso-ventral coordinates

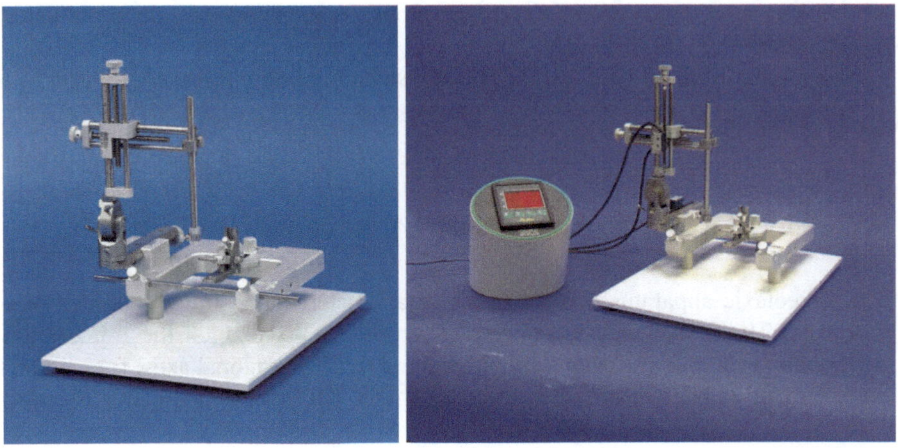

Fig. 4.2 Standard stereotaxic frames from Stoelting®, equipped with a manual (*left photo*) and a digital micromanipulator (*right photo*) for surgery on rats

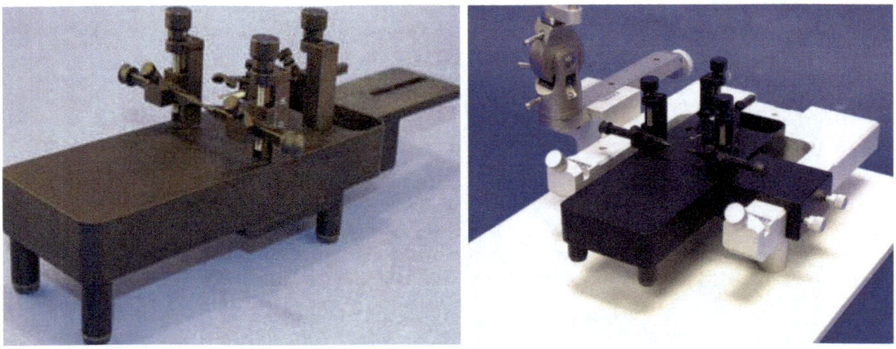

Fig. 4.3 Examples of standard stereotaxic frames from Cunningham®, used alone (*left photo*) or placed in a Stoelting®-type apparatus equipped with a manual micromanipulator (*right photo*) for surgery on mice or small rats

is fixed to the side of the frame and can be moved in three dimensions by three micrometric vernier screw drives. Three points are used to fix the head of the animal in the frame: two laterally adjustable ear bars and a height-adjustable incisor bar.

Commercially available stereotaxic devices are designed for a given species. There are several that exist for the rat, including those provided by ASI Instruments®, Narishige®, Kopf®, and Stoelting® (Fig. 4.2). For the mouse, there are complementary adaptable frames from Supertech Mouse® and Cunningham Mouse® (Fig. 4.3), which can also be used for small rats. In each case, three microdrives allow for a very precise travel in three-dimensional spaces, owing to the three vernier scales mounted on the frame (Fig. 4.4).

In this way, it is possible to move along the antero-posterior "y" axis (AP), which describes a line running from the anterior to the posterior part of the animal's head; the mediolateral "x" axis (ML), which runs from the midline to the right or left side

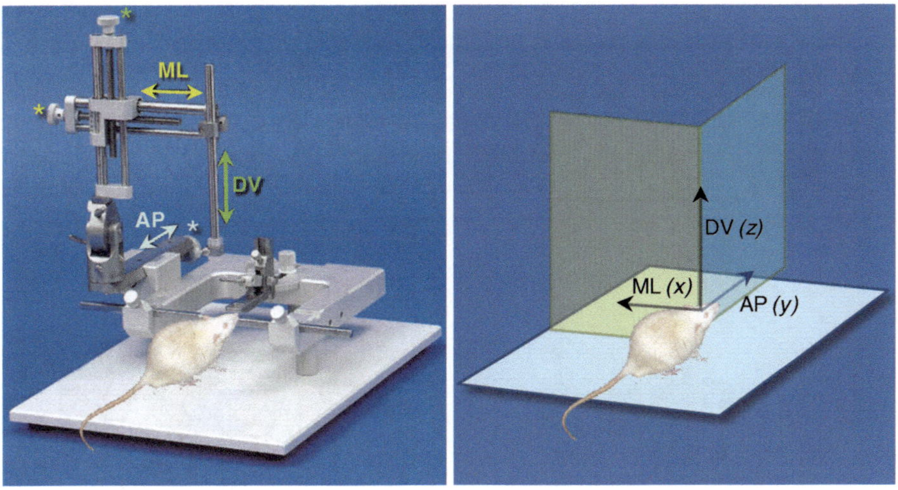

Fig. 4.4 Representation of three micromanipulators in a stereotaxic apparatus (*left figure*) that allow movement in a three-dimensional space represented by the x, y, and z axes (*right figure*). *AP* antero-posterior, *ML* mediolateral, *DV* dorso-ventral

of the head; and the dorso-ventral "z" axis (DV), which runs from the surface of the skull towards the interior of the brain.

The amplitude of travel of each of the three micromanipulators is typically 80 mm. Each micromanipulator consists of a graduated movable arm which travels along a dovetail slide on the fixed arm of the stereotaxic apparatus. The first part of the vernier consists of a graduated scale extending from 0 to 80 (Fig. 4.5) situated on the movable arm. The second portion of the vernier, situated on the fixed part of the arm, consists of a "window" with graduations, from 0 to 10, set 0.9 mm apart. The "0" mark, or principal graduation mark, gives the coordinate reading to the nearest millimeter; the secondary graduation mark on the vernier that lines up with one of the marks on the fixed ruler then gives the reading in tenths of a millimeter.

In this way, the operation of the vernier establishes with precision the distance traveled in one direction or the other. The method of reading the measures on each vernier on the AP, ML, and DV axes is the same.

4.2.1 Stereotaxic Precision: Description of How the Vernier Works

Most stereotaxic devices have an accuracy of 100 μm obtained by means of a vernier. Invented by the mathematician Pierre Vernier in the seventeenth century, the vernier can improve the precision of various analogic measuring instruments, such as a caliper, a device used to measure the distance between two opposite sides of an object, and the goniometer, which measures an angle or allows an object to be rotated to a precise angular position. One can find different types of verniers on stereotaxic devices.

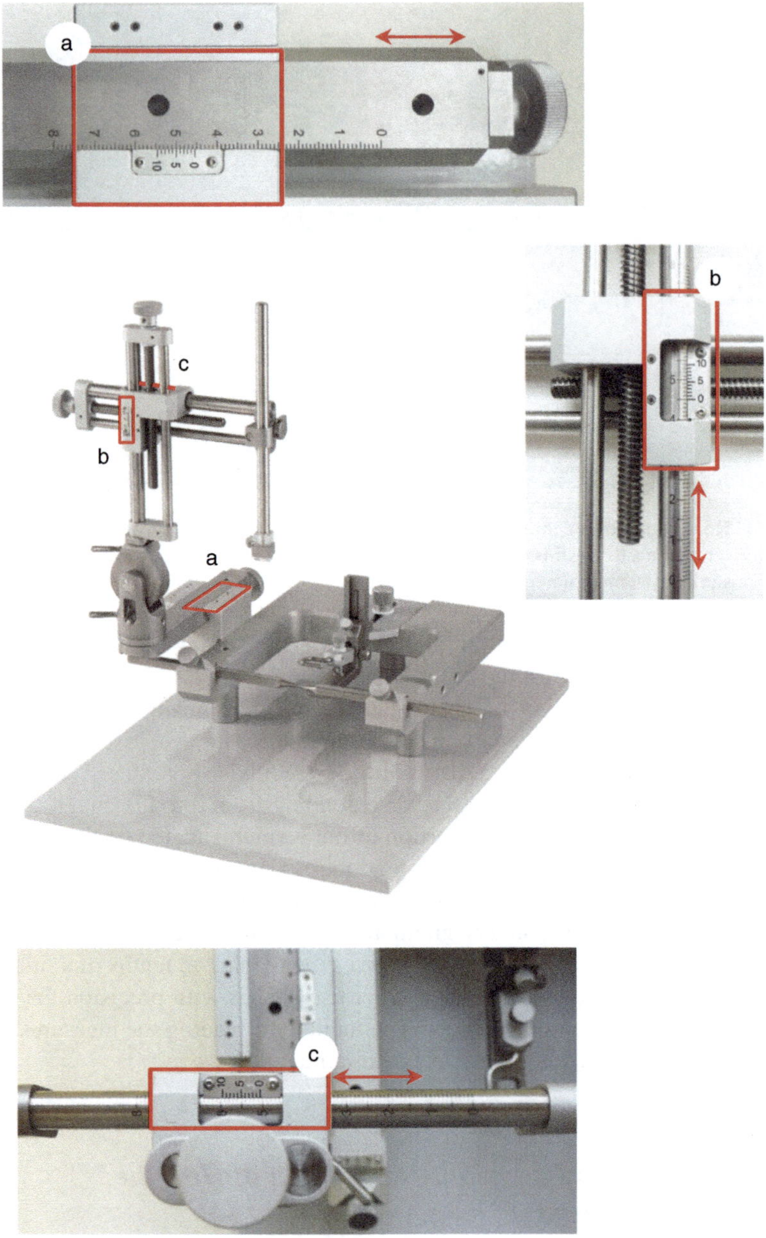

Fig. 4.5 Stereotaxic apparatus and details of three verniers located on the antero-posterior (**a**), mediolateral (**b**), and dorso-ventral (**c**) micromanipulators. A good working knowledge of the vernier is essential for accurately determining the coordinates of the point that you want to target in space. The end of each arm is equipped with a set screw that allows forward or backward transit according to the indications on the vernier

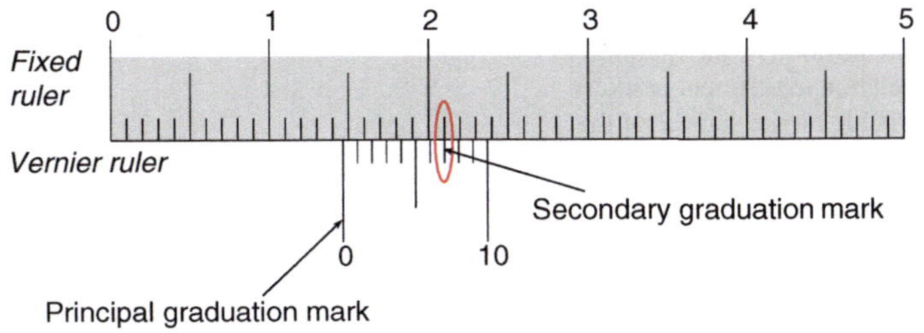

Fig. 4.6 Variant 1

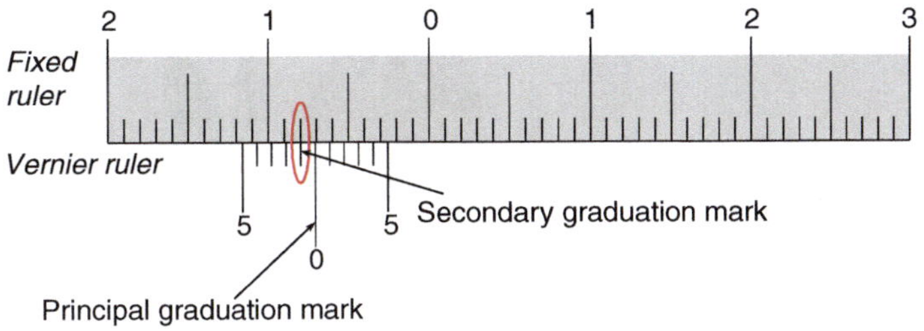

Fig. 4.7 Variant 2

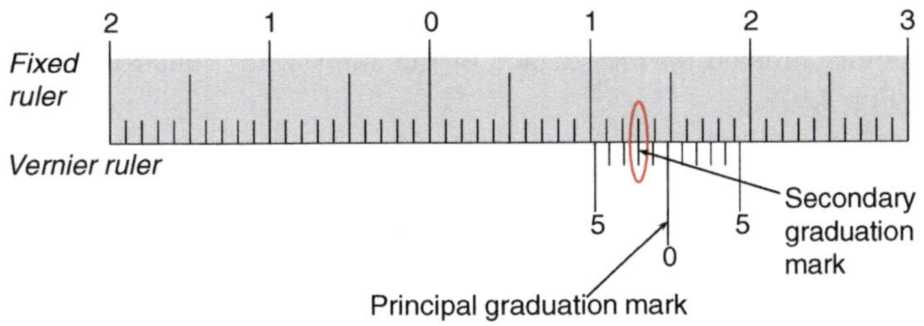

Fig. 4.8 Variant 3

In variant 1 in Fig. 4.6, the reading shows 14.7 mm, that is, a point between 14 and 15 mm on the fixed ruler as read by the principal graduation or "0" mark and 7/10 of a millimeter according to the secondary graduation that aligns with one of the graduations on the fixed ruler. A vernier can move along this graduated rule in both directions. Reading is performed in the same way as described above, first with the principal graduation mark then with the secondary graduation mark. The reading of the vernier scale follows the direction of the fixed ruler.

In variant 2 (Fig. 4.7), the reading gives -7.1 ± 0.1 mm (towards the left).

In variant 3 (Fig. 4.8), the reading gives $+14.8 \pm 0.1$ mm.

4.2.2 *Acquiring the Spatial Coordinates of a Target*

4.2.2.1 Identification of the Cranial Landmarks

Stereotaxic coordinates are used to locate a virtual point in the antero-posterior, mediolateral, and dorso-ventral planes. For a targeted structure, they are calculated

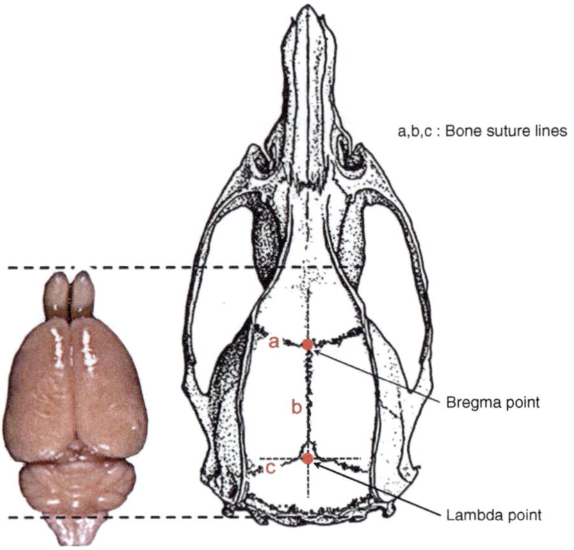

Fig. 4.9 Diagram showing the position of the rat brain in relation to different markers visible on the surface of the skull

relative to coordinates recorded at two reference points: the *bregma* and *lambda* points situated on the surface of the skull at the junction of different bone plates (Fig. 4.9).

4.2.2.2 Determination of Stereotaxic Coordinates

As described above, the coordinates of a target are determined on three axes of orientation: antero-posterior (AP), mediolateral (ML), and dorso-ventral (DV). These are shown in the stereotaxic atlas relative to particular anatomical reference points or landmarks: bregma and lambda (or interaural line). It is up to the operator to choose the appropriate reference point based on the location of the target structure. Typically, the most appropriate point is that which is closest to the target. For example, if it is the hippocampus, a structure located in anterior part of the brain, bregma is recommended as a landmark. On the other hand, if the target is the entorhinal cortex, a structure located in the posterior brain, it is more judicious to use lambda as the reference point.

Note: Calculating the target structure stereotaxic coordinates must take into account the organization of the cerebral vascular system of the animal. The implantation of hardware such as guide cannulae, electrodes, microinjection cannulae, and microdialysis probes must incorporate this topographical information, to avoid as much as possible vascular structures that could be a source of bleeding at the time of insertion (Figs. 4.10 and 4.11, from Dashwood et al. 2010). Thus, depending on the location of the target structure, the cerebrovascular anatomy of the animal can be a constraint in the orientation of the axis of insertion.

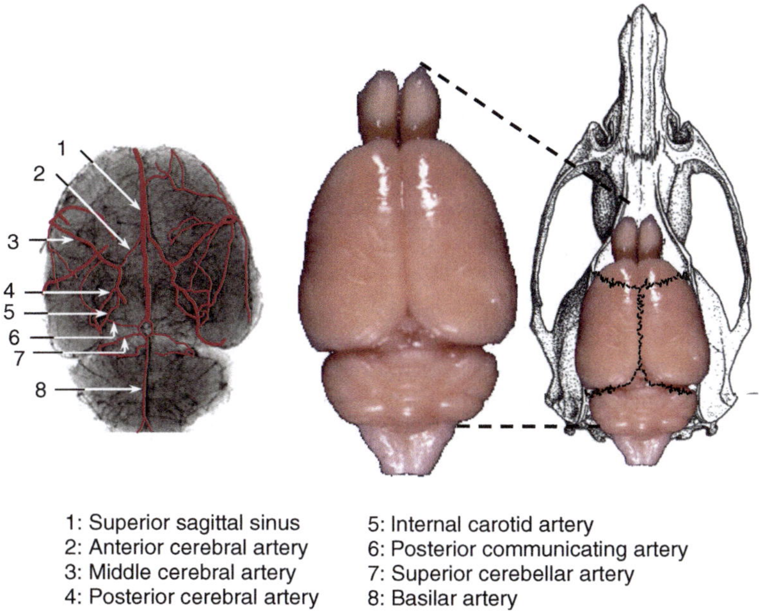

Fig. 4.10 Cartography of the cerebral vasculature in a dorsal view of the rat brain. *Left*: angiography using barium sulfate as imaging contrast. The photo shows vessels at both the dorsal and ventral levels of the brain. For a better visual appreciation, the position of the cranial bone sutures and the brain in situ are shown in the diagrams at *right*

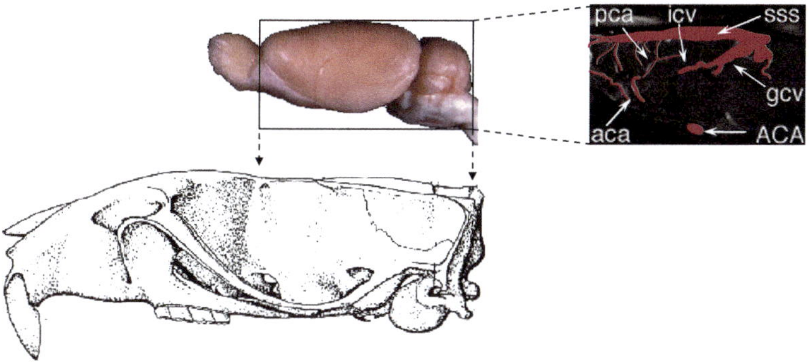

Fig. 4.11 Dorso-ventral cartography of the cerebral vasculature in a sagittal view of the rat brain. *pca* pericallosal artery, *aca* anterior cerebral artery, *ACA* anterior communicating artery, *icv* internal cerebral vein, *gcv* great cerebral vein, *sss* superior sagittal sinus

4.2.3 Example of Determination of Stereotaxic Coordinates: Substantia Nigra Pars Reticulata

After locating the name of the target structure and its abbreviation in the alphabetical list of the stereotaxic atlas (Sect. 4.1.1. p. 70), the corresponding figure is located in the atlas. In our example, the target structure is the *substantia nigra, reticular part*, which corresponds to the abbreviation "SNR" found in Figs. 38–45 in the atlas of Paxinos and Watson (2004). Depending on the region that the investigator wants to target, one of the figures is chosen. This choice is subjective; in our example, Fig. 43 (Fig. 4.12) was chosen from among the 8 other figures since the surface area of the target structure is the largest at this level.

On this figure (below) are represented the structures that can be seen in a coronal section of rat brain taken at a level 5.8 mm posterior to bregma. The point of intersection of the red vertical lines and the blue dotted horizontal line at the center of the target structure indicates that it is 2 mm on either side of bregma and 8.2 mm deep. The estimated three-dimensional coordinates of the center of the structure target are thus AP = −5.8; ML = ±2.0; and DV = −8.2.

The indication "bregma −5.8 mm" shown in Fig. 43 (Fig. 4.12) of the atlas means that the structure is 5.8 mm posterior to bregma and that 5.8 should be subtracted from the coordinate acquired at bregma to arrive on the frontal axis of the target structure. The "±" sign in front of the value 2 indicates that the center of the target structure is 2 mm on either side of the midline; this value should be added to and subtracted from the ML coordinate recorded at bregma in order to be on the sagittal axis of the target structure.

Finally, the negative sign in front of the value 8.2 mm means that the center of the target structure is 8.2 mm below the surface of the skull. In practice, once positioned at the coordinates AP = −5.8 and ML = ±2, 8.2 is subtracted from the value measured on the surface of the skull to attain the center of the target structure.

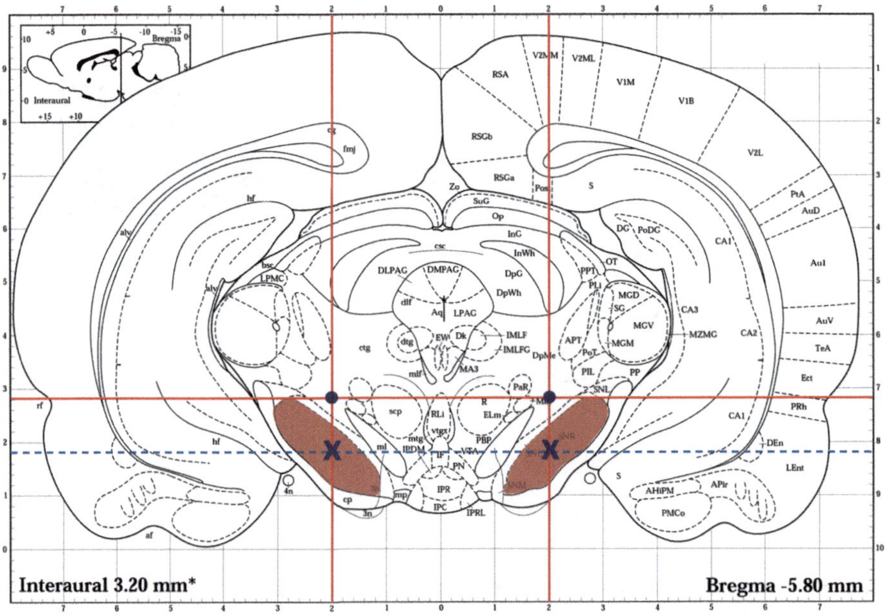

Fig. 4.12 Schematic drawing showing the different brain structures observed in a frontal section of the Wistar rat brain (From Paxinos and Watson 2004). *The substantia nigra pars reticulata* is shown as the *red-colored areas*. "X" indicates the virtual center(s) of the target structure. The *dot* indicates the position of a guide cannula tip that will be implanted 1 mm above the target structure

If a guide cannula is to be placed 1 mm above the target, in order to avoid damaging the latter before a microinjection, the DV coordinate should be adjusted accordingly (see Fig. 4.12). Thus, the final coordinates become AP=−5.8; ML=+2; and DV=−7.2. If the injection cannula is 1 mm longer than the guide cannula, its final point of insertion will then be precisely positioned at the center of the structure (see Fig. 4.12).

4.2.4 *Examples of Determining the Stereotaxic Coordinates of the Vertical Limb of the Diagonal Band of Broca*

In this example, the structure of interest is the *vertical diagonal band* (VDB), as shown in Figs. 13–17 in the atlas of Paxinos and Watson (2004). One of these figures is chosen as a reference for determining the coordinates of the target structure – the most appropriate figure is a subjective choice, depending on the shape and size of the particular structure of interest. In this example, Fig. 15 was selected (Fig. 4.13) since the surface area of the VDB is the largest in this figure. Note that, unlike the previous example of the substantia nigra, the VDB is situated on the midline of the brain.

The coordinates of the target identified using Fig. 15 of the stereotaxic atlas are:

$$AP = +0.7,\ ML = 0,\ \text{and}\ DV = -7.4.$$

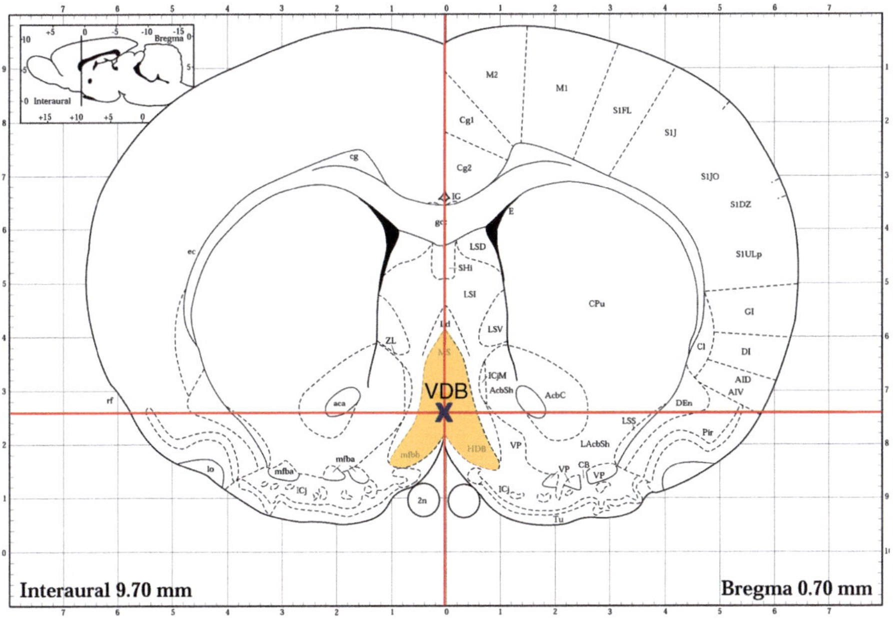

Fig. 4.13 Schematic drawing showing the different brain structures observed in a frontal section of the Wistar rat brain (From Paxinos and Watson 2004). The vertical part of the diagonal band of Broca is shown as the *yellow-colored area*. "X" indicates the virtual center of the target structure

The indication “bregma 0.70 mm” shown in the figure of the atlas means that the target structure is situated 0.7 mm in front of bregma; thus, 0.7 mm has to be added to the coordinate measured at bregma to be on the frontal axis of the target structure. The plot of the vertical line passing through the center of the target structure indicates that the target is vertically aligned with the bregma point. The ML value=0 means that the center of the structure is in the sagittal midline of the brain. The negative sign in front of the value 7.4 means that the target coordinate is located 7.4 mm below the surface of the skull.

Note: If one refers to the cartography of the vascular system (see Figs. 4.10 and 4.11), it will be evident that a lateral approach to the target will be necessary for avoiding the sagittal venous sinus. For this, it suffices to calculate an angle at which the structure can be approached. The stereotaxic instrument allows for the precise adjustment of the inclination of the dorso-ventral manipulator arm and object carrier, in order to approach the target structure at a defined angle. To facilitate the calculations in this example, a 10° angle, largely sufficient to avoid the vascular obstacle, has been chosen (Fig. 4.14)

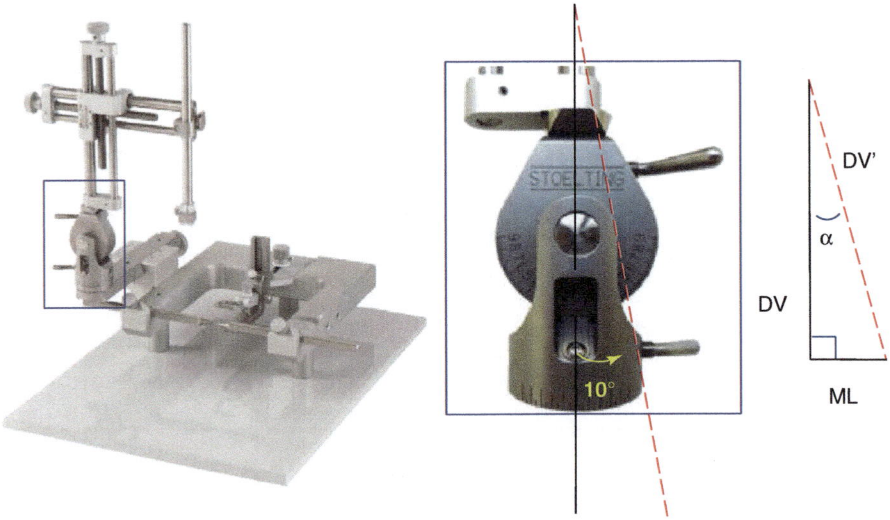

Fig. 4.14 Enlargement of a photograph of the dorso-ventral micromanipulator. The micromanipulator can be inclined precisely as shown in the *center photo*, at an angle chosen by the operator, for example, 10°, as shown in the diagram on the *right*

- For the calculation of the coordinate DV′ for implanting the cannula with an angle alpha of 10°, we know that:
  Cosinus *alpha* = DV/D′, Cos 10 = 0.9848, and DV = 7.4 according to the atlas; thus, DV′ = 7.4/0.9848 = 7.5142 or 7.51 mm.
- To calculate the ML' coordinate secondary to angle *alpha,* we know that: Sinus *alpha* = ML′/DV′; thus, ML′ = Sin *alpha* × DV′ = 0.1736 × 7.51 or 1.3 mm.

The final coordinates will thus be:

$$AP = +0.7,\ ML' = \pm 1.3,\ \text{and}\ DV' = -7.51.$$

As in the previous example, the ML coordinate is associated with a ± sign because it is possible to approach the center of the target structure by inserting the cannula at a 10° angle from either the right or left side of midline (Fig. 4.15).

In practice, the angle needs to be adjusted at the base of the dorso-ventral manipulator arm, in order to properly direct the descent of the object carrier (Fig. 4.16).

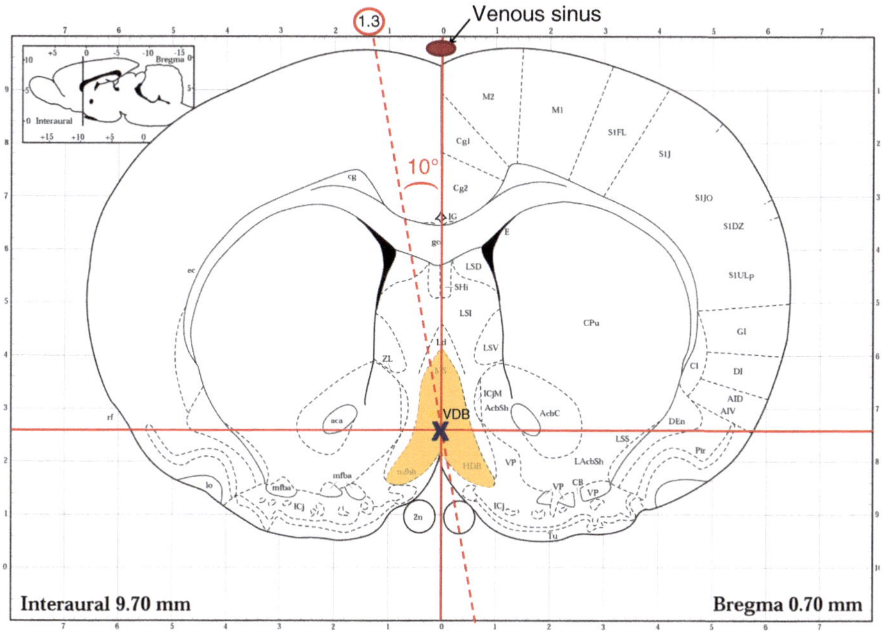

Fig. 4.15 Diagram showing the different structures observed in a frontal section of the Wistar rat brain (Fig. 15 in the atlas of Paxinos and Watson 2004). The *red dotted line* represents an ideal approach to the target structure, avoiding the vascularization and particularly the venous sinus. The vertical part of the diagonal band of Broca is shown as the *yellow-colored area*. "X" indicates the virtual center of the target structure

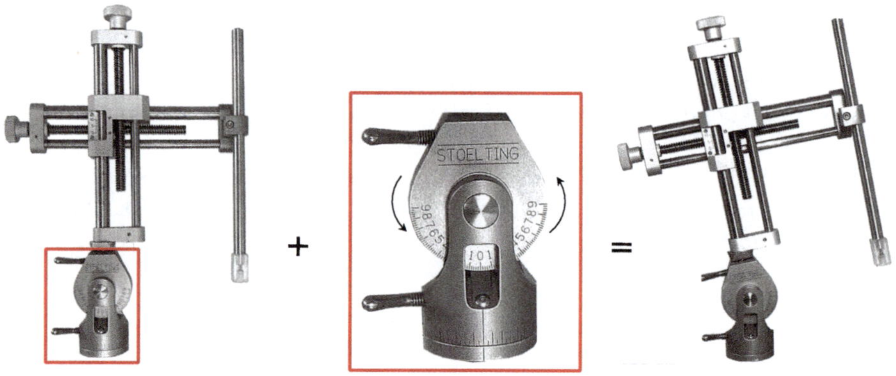

Fig. 4.16 Photographs of the dorso-ventral manipulator arm, before and after applying an angle of 10°

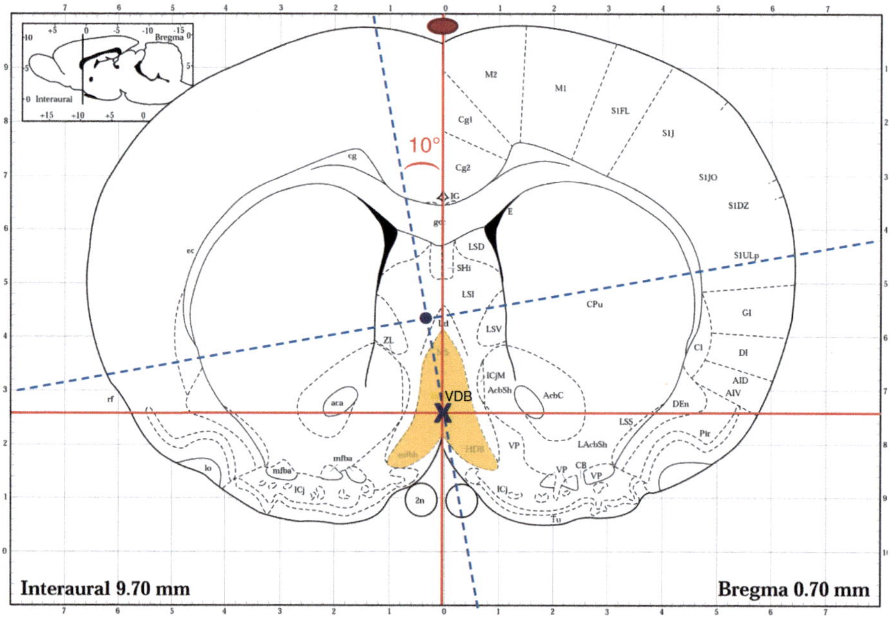

Fig. 4.17 Diagram showing the trajectory of the guide cannula and its injection cannula, in the case where the guide cannula is to be placed 2 mm above the target structure. The *horizontal dotted line* represents the plane perpendicular to the path of descent of the cannula. "X" represents the virtual center of the target structure. The *dot* represented at the intersection of the *dotted lines* represents the position of the tip of the guide cannula 2 mm above the target structure

If a guide cannula is to be placed 2 mm above the structure, the final coordinates, shown in Fig. 4.17, will be:

$$AP = +0.7,\ ML' = \pm 1.3,\ \text{and}\ DV = -5.51.$$

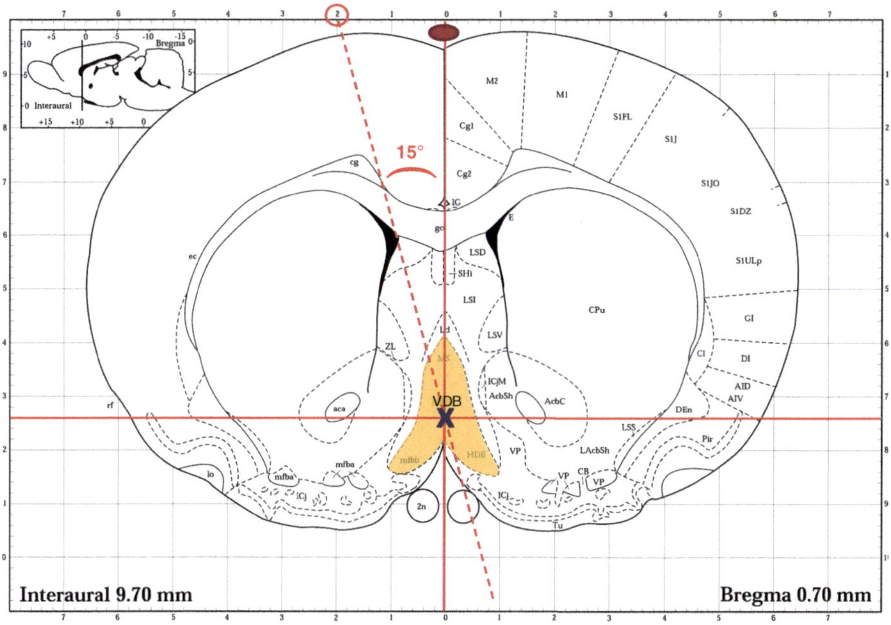

Fig. 4.18 Diagram showing the different structures observed in a frontal section of the Wistar rat brain (Fig. 15 in the atlas of Paxinos and Watson 2004) at a distance 0.7 mm anterior to bregma. The *dashed red line* represents the trajectory calculated for reaching the target structure at an angle. The vertical part of the diagonal band of Broca is represented by *yellow-colored area*. "X" marks the virtual center of the target structure

Another possibility is to arbitrarily choose an integer value for the ML coordinate (here 2 mm) and then calculate the angle of inclination for avoiding the venous sinus. To calculate the angle *alpha* for a ML coordinate = 2 mm, we know that:

Tan *alpha* = ML/DV, DV = −7.4, and ML = 2; thus, tan *alpha* = 2/7.4 = 0.27, *alpha* = 15.1° or 15°.

For the calculation of the DV coordinate secondary to angle *alpha*, we know that:

Sinus *alpha* = ML/DV′, ML = 2 mm and *alpha* = 15°, and DV′ = ML/sinus *alpha* = 2/0.259 = 7.72.

Thus, the final coordinates (Fig. 4.18) are:

$$AP = +0.7,\ ML = \pm 2, \text{and}\ DV = -7.72.$$

As in the preceding examples, if a guide cannula is to be placed 2 mm above the target structure, the final coordinates (Fig. 4.19) are:

$$AP = +0.7,\ ML = \pm 2,\ \text{and}\ DV = -5.72.$$

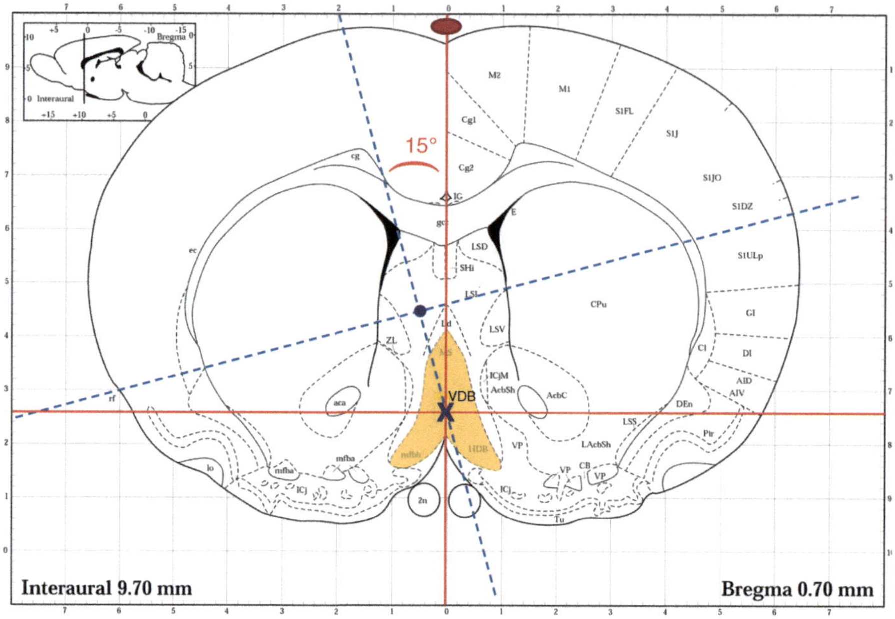

Fig. 4.19 Diagram showing the trajectory of the guide cannula and its injection cannula, in the case where the guide cannula is to be placed 2 mm above the target structure. The *horizontal dotted line* represents the plane perpendicular to the path of descent of the cannula. "X" represents the virtual center of the target structure. The *dot* represented at the intersection of the two dotted lines marks the position of the tip of the guide cannula, 2 mm above the target structure

Note: A very rigorous approach is necessary for the correct determination of the dorso-ventral coordinate. Some publications report a determination of the dorso-ventral coordinate either from the surface of the skull or from the surface of the dura mater. One must be vigilant about the starting point of the descent of the cannula, since there can indeed be a difference of up to 1 mm between those two reference points. In addition, the convex shape of the cranial bone is important to consider in determining the dorso-ventral coordinate (Fig. 4.20).

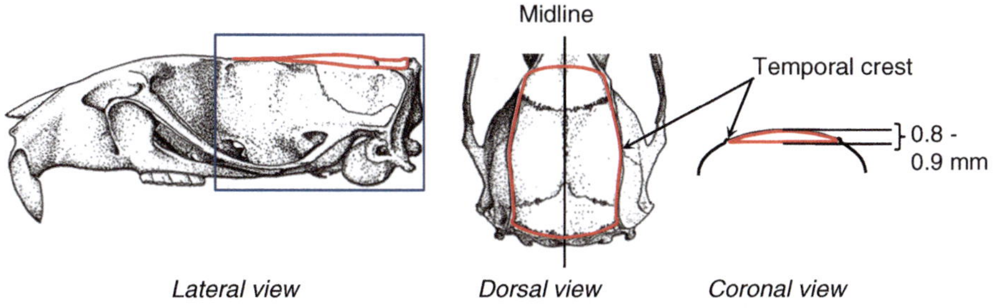

Fig. 4.20 Diagrams representing the surface of the skull in lateral, dorsal, and coronal views. The *red line* indicates an arbitrary work surface. The side (lateral) view shows that the skull surface is convex and that a gap of 0.8–0.9 mm can exist between the highest point on the midline and the lowest point on the lateral edge of the parietal bone (the bone "stop") shown in the coronal view

> *Note*: The bregma and lambda points are located on the transverse axis (midline) of the skull. The DV coordinate of the target structure determined from one of these two points will be valid *only* if the target is located vertically in line with one of these points. If this is not the case, a descent has to start from the AP and ML coordinates acquired directly above the target, measuring the DV coordinate from here to the final coordinate that has been calculated for the target structure.

References

Albe-Fessard D, Stutinsky F, Libouan S (1966) Atlas stéréotaxique du diencéphale du rat blanc. Éditions du Centre National de la Recherche Scientifique, Paris

Altman J, Bayer SA (1994) Atlas of prenatal rat brain development. CRC Press, Inc., Boca Raton

Cooley RK (1990) Stereotaxic surgery in the rat: a photographic series. A.J. Kirby Co, London

Dashwood MR, Loesch A (2010) Endothelin-1 as a neuropeptide: neurotransmitter or vascular effects? J Cell Commun Signal 4(1):51–62

De La Porte C (1993) Technical possibilities and limitations of stereotaxy. Acta Neurochir (Wien) 124(1):3–6

Franklin KBJ, Paxinos G (2008) The mouse brain in stereotaxic coordinates, 3rd edn. Academic/Elsevier, Oxford

König JFR, Klippel RA (1963) The rat brain: a stereotaxic atlas of the forebrain and the lower parts of the brain stem. Williams and Wilkins, Baltimore

Lehmann A, Gautier M, Ghilini G, Henry J, Langlois R, Laplante S (1974) Atlas stéréotaxique du cerveau de la souris. Éditions du Centre de la Recherche Scientifique, Paris

Montemurro DG, Dukelow RH (1971) A stereotaxic atlas of the diencephalon and related structures of the mouse. Mount Kisco/Future, New York

Paxinos G, Franklin KBJ (2001) The mouse brain in stereotaxic coordinates, 2nd edn. Elsevier, New York

Paxinos G, Franklin KBJ (2003) The mouse brain in stereotaxic coordinates: compact, Compact 2nd edn. Elsevier, New York

Paxinos G, Franklin KBJ (2012) The mouse brain in stereotaxic coordinates, Compact 4th edn. Elsevier, New York

Paxinos G, Watson C (1998) The rat brain in stereotaxic coordinates, Compact 4th edn. Academic, New York
Paxinos G, Watson C (2004) The rat brain in stereotaxic coordinates, 5th edn. Academic, San Diego
Paxinos G, Watson C (2007) The rat brain in stereotaxic coordinates, 6th edn. Academic, San Diego
Paxinos G, Watson C (2013) The rat brain in stereotaxic coordinates, 7th edn. Academic, San Diego. ISBN 9780123919496
Pellegrino LJ, Cushman AJ (1967) A stereotaxic atlas of the rat brain. Appleton-Century-Crofts, New York
Schambra UB, Lauder JM, Silver J (1991) Atlas of the prenatal mouse brain. Academic, New York
Sidman RL, Angevine JB Jr, Taber PE (1971) Atlas of the mouse brain and spinal cord. Harvard University Press, Cambridge. Mass
Swanson L (2003) Brain maps: structure of the rat brain, 3rd edn. Elsevier, New York

Chapter 5
Preparation of the Stereotaxic Surgical Procedure

5.1 General Preoperatory Conditions

5.1.1 Assessing the General Condition of Animals on the Day Before and on the Day of Surgery

Before any surgical procedure, it is important to ensure that the animals are in good health by checking their appearance and analyzing their general behavior. A clinical examination may be required to detect certain anomalies and to bring these to the attention of the investigator. In a neurosurgical procedure, it is necessary to eliminate any animal with an abnormality that can cause anatomical or behavioral alterations and thus influence the experimental results (Appendix 1). Signs of reduced appetite; weight loss; deficits in normal exploratory behavior; hyperresponsiveness to handling; vocalization; self-mutilation; prostration; bite marks; scratches; patchy, dull, and/or ruffled fur; and unusual posture or facial expression should immediately exclude the animal from the procedure. Any of these signs (Appendix 3) should be noted on the animal's follow-up register (Appendix 2B) and reported to the veterinarian to determine the corrective course of action, treatment, prophylactic measures, or eventual euthanasia (Appendix 4).

5.1.2 Preparation of Surgical Equipment

The investigator responsible for the experiment and the surgical procedure must ensure the availability of the required equipment and its maintenance, preparation, and sterilization. A sufficient amount of time must be given to do all of this in advance, so as not to be caught short at the last moment (Appendix 5). The availability of the required anesthetics and their expiration dates should be verified. The laboratory stock and the administration of any anesthetic should be recorded in a

B. Ferry et al., *Stereotaxic Neurosurgery in Laboratory Rodent*,
DOI 10.1007/978-2-8178-0472-9_5, © Springer-Verlag France 2014

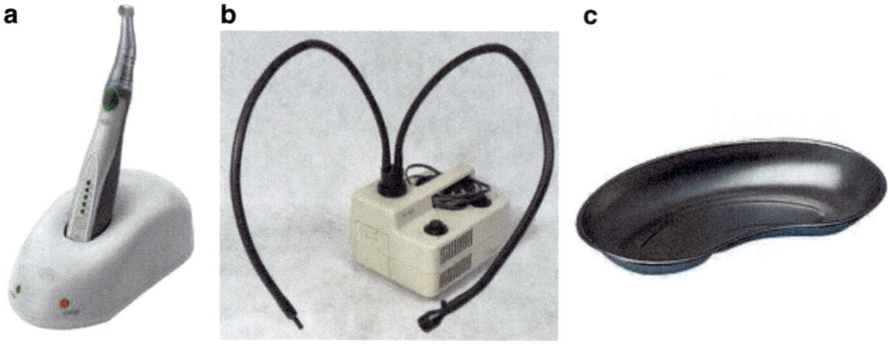

Fig. 5.1 Benchtop instruments. (**a**) Dental drill with angled handpiece, for piercing the skull bone at the time of the insertion procedure. Inserted material (cannula, electrode, etc.) must be chemically sterilized prior to the procedure. (**b**) Cold light source (double gooseneck fiber optic lamp). (**c**) Surgical kidney dish for disposal of organic waste

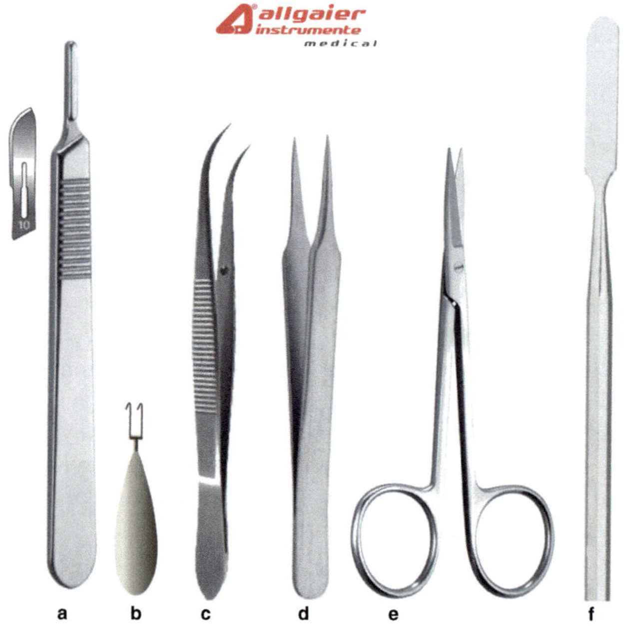

Fig. 5.2 Surgical instruments necessary for the incision procedure. (**a**) Scalpel handle and blade for skin incision, (**b**) Ferry tissue retractor prototype that will be manufactured by ALLGAIER Instrumente Medical. (**c**), (**d**) Straight and curved Dumont forceps for gripping the tissue. They are also used for grasping the microscrews and implantation materials. Excessive force should not be applied to avoid deforming the forceps jaws. (**e**) Fine surgical scissors. (**f**) Spatula for the preparation of dental cement

register. This register is maintained by the laboratory personnel in charge of the pharmacy service within the institution of experimentation.

Benchtop instruments, including a cold light source and dental drill handpiece, are placed on the operating table next to the stereotaxic apparatus, along with a receptacle, such as a surgical kidney dish or plastic container, for organic waste (Figs. 5.1 and 5.7). A surgical box containing the sterilized surgical instruments (Figs. 5.2 and 5.7) is placed on a trolley in the clean zone (Fig. 5.8). From our

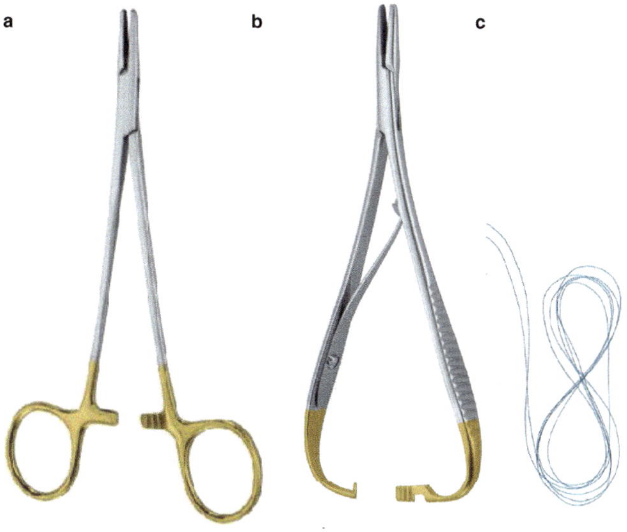

Fig. 5.3 Suturing instruments. (**a**) Standard needle holder ALLGAIER Instrumente Medical (**b**) Mathieu-type needle holder ALLGAIER Instrumente Medical. (**c**) Suture thread with attached needle as found in its sterile blister package

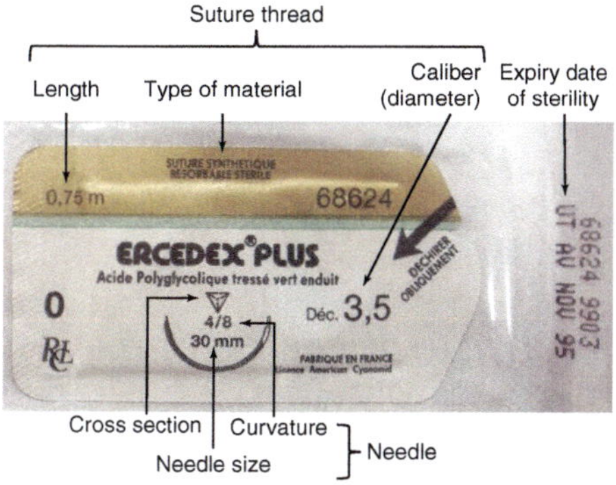

Fig. 5.4 Example of a commercially available prepackaged suture

experience, it is necessary to have a minimum of instruments such as those shown in Fig. 5.2 in order to carry out an incision of the head. A surgical box containing the suturing instruments (Figs. 5.3 and 5.7) is kept on a sterile surgical tray pending their use at the end of the surgery.

The rigorous undertaking of a surgical procedure requires the an assistant for the successive opening of different storage boxes containing the surgical drapes, instruments, disposable materials, and suture blister packs so that the gloved surgeon can access them without breaking the chain of asepsis. Sterilized disposable materials, cotton swabs, needles, and sutures (Fig. 5.4) should be placed on the sterile tray next to the box containing the surgical instruments. A sterile drum containing the operating drapes and gauze are placed nearby on a trolley in the "clean" zone (see Fig. 5.8). Surgical caps, masks, gowns, and gloves are made accessible to the operator to facilitate an aseptic procedure. Tips on the selection of suture material are provided in Fig. 5.4 and Table 5.1.

Table 5.1 Numbering codes and corresponding diameters of various suture threads

Threads for experimental surgery (rodents)						
	Threads of microsurgery			Threads of general surgery		
Caliber (diameter) n° *Metric n°*	0.4	0.5	0.7	1	1.5	2
Commercial n° *UPS (US pharmacopoeia) n°*	8/0	7/0	6/0	5/0	4/0	3/0
Diameter in microns	45	60	85	125	175	225

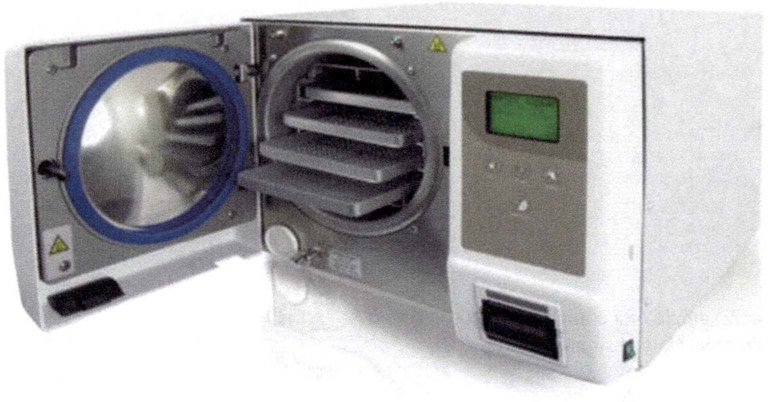

Fig. 5.5 Benchtop autoclave (18 l capacity, Webstore Medical®) with a compact footprint (25 × 45 cm)

5.1.3 Sterilization of Surgical Equipment

Complications of infection related to a surgical procedure is a permanent risk in animal surgery. Sterilization of the surgical equipment is the first element in preventing postoperative infection. Two sterilization methods are commonly used.

- *Heat sterilization,* since beyond a certain temperature, all life is impossible. Microorganisms, such as heat-resistant bacterial spores, are destroyed by heating for 30 min at 170 °C. This method can be used for heat-resistant material, including the implant materials (cannulae, electrodes, obturators) and the special equipment and tools used for drilling the skull bone. It can be performed using either an autoclave (Fig. 5.5), which ensures stability and a traceability of a recorded temperature curve, or even a domestic oven, provided that the heating conditions are increased (45 min at 170–180 °C). Cloth surgical drapes, gowns, and compresses can also be heat-sterilized in a drum (Fig. 5.6).
- *Sterilization by immersion in an antiseptic solution* can be used as an emergency workaround option. The material is immersed in an antiseptic solution for a length of time recommended by the manufacturer, usually around 10 min. Glutaraldehyde-based products containing corrosion inhibitors (Korsolex®, IS Mucadont®) offer good results, provided that the instruments are rinsed with sterile water before use.

Fig. 5.6 Examples of sterile drums (Covéto®)

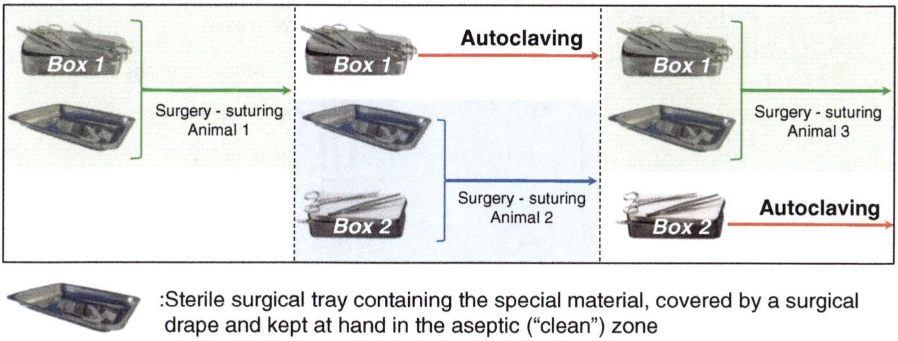

Fig. 5.7 Example of using two boxes of instruments during the same surgical procedure

It is important to separate the specialized (and often expensive) equipment from the surgical instruments and materials needed at hand during the operation. *Two sets of instruments* for the incision and suturing (scalpel, dissecting forceps, drill bits, scissors, and needle holder) allow the cleaning and sterilization of the first set used on animal "X," while the second set is being used on animal "X+1." Special equipment that usually has little or no contact with the tissues of the anesthetized animal can be kept sterile by placing it in a tray covered with a sterile drape between each procedure (Fig. 5.7). The ear and incisor bars can be autoclaved or cleaned simply with a disinfectant towelette.

5.1.4 *Organizing and Managing Surgical Asepsis*

A coherent organization of a surgical room, operating theater or a lab benchtop space used for surgery, is crucial to carrying out sterile surgery and limiting the risk of breaking the chain of asepsis. The principle of an operational workflow (a forward moving sequence of steps) commonly used in industrial sectors helps in organizing the different levels of cleanliness or sterility required at each stage of the procedure (Fig. 5.8). This organization identifies a "dirty" preparation zone in which *shearing of the animal* and *induction of anesthesia* is carried out and an "intermediate" zone where the paws of the animal are washed with antiseptic soap

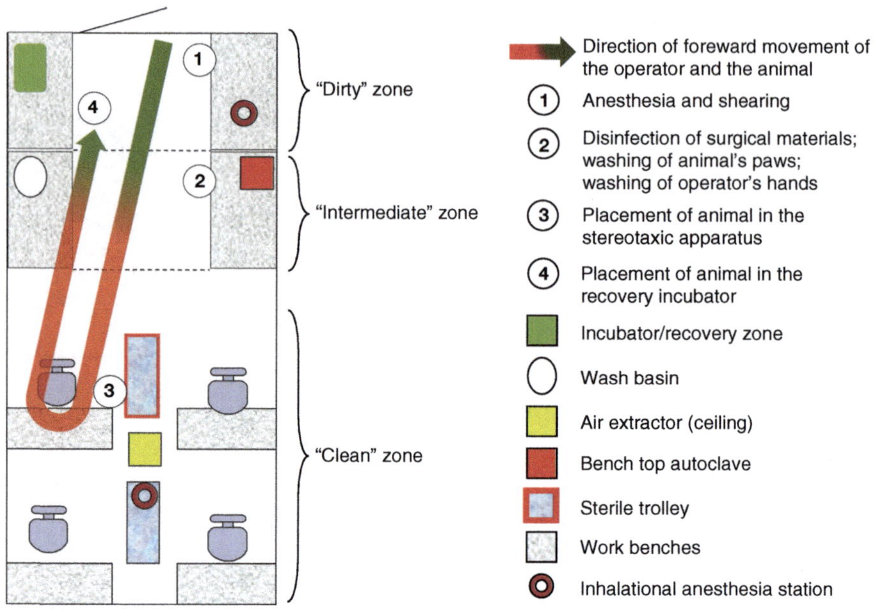

Fig. 5.8 Organization of a surgery room. The *arrow* represents the forward moving sequence of steps in the surgical procedure

before bringing it into the "clean" zone on a thermostated electric heating blanket or pad equipped with a *rectal temperature probe*, just prior to being *fixed in the stereotaxic apparatus.*

After the disinfection of the top of the head (with soap, rinsing, 70 % ethanol, and a disinfectant such as chlorhexidine or polyvinylpyrrolidone iodine[1]), a subcutaneous injection of *local anesthetic* (see Sects. 5.2.3 and 5.2.4) is made on each side of the line of incision. This helps to limit the postoperative pain of the animal. Precision in carrying out the surgical procedure is important for limiting pain, morbidity, and mortality, hence, the importance of proper training in neurosurgical practices.

The first package of sutures opened by the assistant is put it on the surgical tray so the surgeon can take the sterile part of the package and use it at the end of the surgery for the suture.

Each micrometer screw of the stereotaxic apparatus is then disinfected, as well as the handpiece of the dental drill. With the aid of an assistant, the operator then dons a gown and mask from the sterilization drum (see Fig. 5.6), as well as sterile gloves. At this point, the operator proceeds with the set up of the sterile zone with help from the assistant who passes to him the different open boxes of materials. A sterile drape is spread to create the operating field; the different surgical instruments, compresses, and other consumables, as well as any drapes required for the animal, are then successively placed upon the drape. The first package of sutures

[1] On average, there are ten million bacteria per square centimeter of skin. A simple washing with soap can destroy 99.9 %. This still leaves more than 10,000 bacteria per cm^2.

opened by the assistant is put on the surgical tray so the surgeon can take the sterile part of the package and use it at the end of the surgery for the suture.

During successive procedures, equipment and materials are covered with a sterile drape while the second animal is being prepared and then reused once the operator has changed gloves. Before the skin incision, the operator makes a fresh disinfection of the *operating zone*[2] with chlorhexidine or polyvinylpyrrolidone iodine and then puts the surgical drapes in place.

At the end of the surgical procedure, the animal is placed in the recovery zone, for example, in an incubator or under a heating lamp in the preparation zone. This allows the operator to proceed, if necessary, with an operation on a second animal while monitoring the recovery of the first.

Ideally, all soiled instruments are cleaned or left to soak in the appropriate product in the "intermediate" zone. Once rinsed and dried, they can be autoclaved before reuse. For the second animal, the operator proceeds as before with the shearing and anesthesia in the preparation zone and then the setup of the animal. Normally, the entire procedure of dressing is repeated, in particular a change of surgical gowns, but it is common to limit this to a simple change of gloves.

5.2 Pain and Its Pre-, Intra-, and Postoperative Treatment

5.2.1 Definition

According to the International Association for the Study of Pain, pain is an "unpleasant sensory or emotional experience caused by actual or potential tissue injury… that provokes protective motor and autonomic reactions leading to a change in the behavior specific to the individual. The inability to communicate does not exclude the possibility that the subject feels pain and requires adequate analgesic treatment." At the end of the twentieth century, a British survey of 1,000 practicing veterinarians revealed that only half of the animals undergoing an ovarian hysterectomy and two-thirds of those having abdominal surgery received an opioid analgesic or anti-inflammatory drug during their operation. Nearly one surgeon in 10 felt that "the surgical operation in question did not cause enough pain to warrant treatment." Within the last decade, attitudes have changed significantly – the entire scientific community agrees that pain, if left untreated, increases the risk of morbidity and mortality and constitutes a significant bias for experimental results.

Pain in animals is difficult to assess because the subject is unable to directly communicate what it feels. Only those signs that are secondary to painful phenomenon are expressed and can thus be observed and treated retrospectively. It is nevertheless essential to anticipate and consider it and to deal with it more effectively. This is not only a moral issue but also a scientific and regulatory requirement. As a precautionary principle, it is considered that any procedure perceived as being

[2] A renewed disinfection again destroys 99.99 % of bacteria and is considered complete when a threshold survival level of 0.01 % is achieved.

Table 5.2 Different types of stimuli and the pain associated with them

Site	Type of stimulus	Example
Skin	Thermal	Burn
	Mechanical	Cut
	Chemical	Inflammation
Muscle	Mechanical	Blow, compression
	Chemical	Inflammation or intra-articular injection of irritant

painful in humans be regarded as such in animals and that a pretreatment thereof with diverse agents (e.g., opioids, steroidal or nonsteroidal anti-inflammatory drugs, local anesthetics, alpha-2 adrenergic receptor agonists) should be implemented. A review of the anatomical mediators of pain transmission is useful here for understanding the concept of more effective multimodal analgesia, which can allow the use of lower doses of drug and thus a reduction in the incidence of their side effects.

5.2.2 The Physiology of Pain: From Transduction to Perception

Nociception, or pain, is a physiological process that involves five steps: transduction, transmission, modulation, projection, and perception of a painful signal.

Transduction is a process by which nociceptors, specialized free nerve endings that are sensitive to mechanical, chemical, or thermal stimuli, convert these stimuli into signals (Table 5.2) that are relayed by neurons to higher centers. Nociceptors are present in the skin, muscles, joints, bones, and viscera.

Peripheral sensation is transmitted by three types of nerve fibers: A*beta*, A*delta*, and C. Under normal conditions, the A*beta* fibers carry non-painful sensory information such as friction, vibration, pressure, or proprioception. Nociceptors on A*delta* and C fibers generate painful sensory information that is carried by the A*delta* and C fibers to the gray matter in the dorsal horn of the spinal cord. This process is referred to as *transmission* (Fig. 5.9).

Within the spinal cord, a synaptic relay allows for a local *modulation* of the nociceptive signal by the intervention of three types of postsynaptic neurons:

- Modulatory interneurons responsible for local signal modulation.
- Motor neurons responsible for the neural reflex arc.
- Neurons of the ascending pathways that participate in directing the nociceptive signal towards centers of integration in the brain (thalamus, hypothalamus, and reticular formation) by way of the spinothalamic, spinoreticular, or trigeminal tracts. This process is called *projection.* The integration of the pain signal by cortical structures is referred to as *perception.*

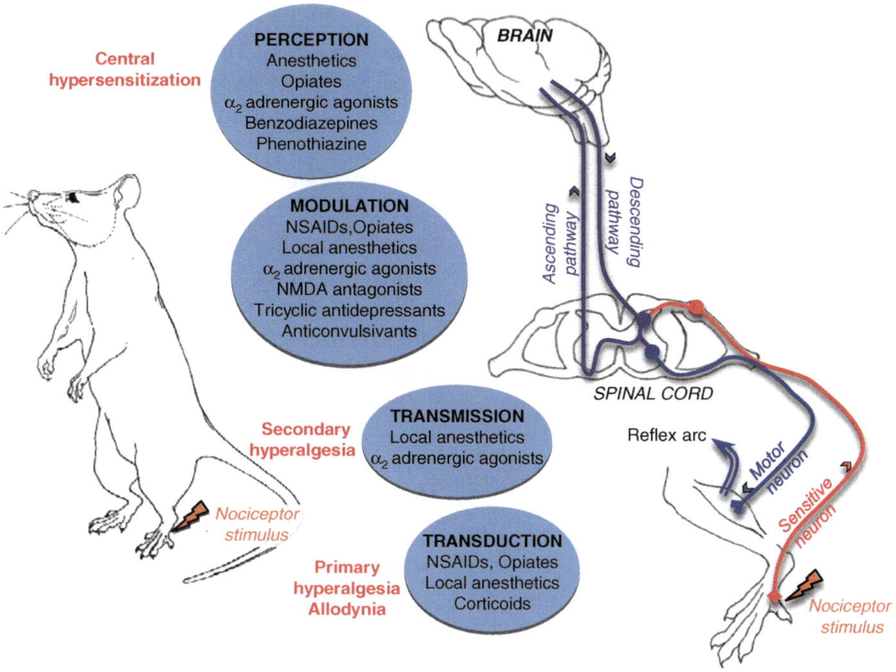

Fig. 5.9 The major pathways of nociceptive information processing

Nociception: a physiological process by which pain signals are routed from pain receptors towards the brain. Nociceptors: pain receptors. Nociceptive stimuli: phenomena capable of activating nociceptors and generating a painful response. NSAIDs : nonsteroidal anti-inflammatory drugs

5.2.3 *Amplification of the Pain Signal by Sensitization*

Regulatory mechanisms are continuously solicited and established by the organism. A nociceptive signal can be amplified by the phenomena of peripheral and central sensitization. While discussed individually here for a better understanding, peripheral and central sensitizations are intricate and inseparable mechanisms (see Fig. 5.9).

- *Peripheral sensitization* is related to a release of endogenous substances, such as inflammatory mediators, potassium, and ATP, into the local environment of the nociceptors in response to tissue damage and/or inflammation. This sensitization provokes a peripheral hyperalgesia, an exaggerated response to a noxious stimulus in the immediate vicinity of the lesion, and allodynia, an increase in the reactivity of nociceptors to stimuli that are not painful under normal conditions.

Peripheral sensitization can be partially treated by the intravenous administration of nonsteroidal anti-inflammatory drugs (NSAIDs). These agents inhibit the production of certain mediatory substances, but other endogenous compounds not directly involved, such as ATP and substance P, continue to have a sensitizing effect.

- *Central sensitization* involves the synaptic relay in the spinal cord. Repeated and intense noxious stimuli cause temporal summation or *windup*. This activates NMDA (*N*-methyl-D-aspartate) receptors and lowers the threshold of excitability of postsynaptic neurons, resulting in a sensitization of regions several centimeters distant from the injured area, otherwise referred to as *central hyperalgesia*. In this context, nerve fibers that usually convey non-painful sensory messages transmit a modified signal that can be interpreted as painful. This is called *allodynia*.

5.2.4 Optimized or Multimodal Analgesia

"Anesthesia or the use of a sedative, even a powerful one, is not synonymous with analgesia" (Smith et al. 1985). The knowledge and understanding of painful phenomena has led to the development of effective strategies for analgesia. Multimodal analgesia aims to act at different stages of pain signal propagation, based on the combined use of compounds from different drug classes and/or different methods of administration, e.g., from intravenous for rapid and widespread onset and action to epidural or local injection for direct action on nerve fibers. These combinations exploit the complementarity and the eventual synergy between agents, permitting the use of lower doses and thus reducing the risks of adverse side effects (Shavit et al. 2005).

The most commonly used agents for blocking conduction in peripheral nociceptive fibers are the local anesthetics (2 % lidocaine and 0.5 % bupivacaine), which act by inhibiting the operation of sodium channels. They can be injected directly along specific nerve fiber projections (e.g., brachial plexus, cranial and intercostal nerves) or at the level of the nerve root(s) (e.g., epidural injection). Opioids, such as morphine, can also be administered by the epidural route.

Commonly used dosages of the major analgesic drugs are presented in Appendix 7.

5.2.4.1 Analgesic and Anti-hyperalgesic Effects of Lidocaine (from Joris 2008)

Lidocaine, administered by local or regional routes, has analgesic properties. Analgesia is exerted by specific blockade of tonic discharges in peripheral neurons evoked by nociceptive stimuli (Tanelian and MacIver 1991) and by increasing the excitation threshold of A∂ and C nerve fibers (Wallace et al. 1996). A direct action on nociceptive transmission at the spinal level has also been described (Nagy and Woolf 1996). Lidocaine blocks sodium channels (Kuno and Matsuura 1982), resulting in a decrease of postsynaptic depolarization triggered by the activation of neurokinin and NMDA receptors and a consequent reduction in the generation of hyperalgesia and allodynia. In animals, secondary nociceptive responses to visceral stimuli, rectal distension, and visceral motor and cardiovascular reflexes are also inhibited by lidocaine. The analgesic effect is dose dependent, attained at plasma concentrations of 2 μg/ml, with low interindividual variability (Ferrante et al. 1996). In addition to its analgesic effect, lidocaine is anti-hyperalgesic (Koppert et al. 2000).

Local anesthetics, NSAIDs, alpha-2 adrenergic receptor agonists (e.g., medetomidine), NMDA receptor antagonists (e.g., ketamine), tricyclic antidepressants, and anticonvulsants have actions at the level of modulation.

Opioids and alpha-2 adrenergic agonists are the major molecules, which, upon intravenous injection, act upon the synaptic relay of pain signals at the level of the dorsal horn in the spinal cord. These agents act on different types of receptors in the pre- and postsynaptic zones. When used in combination, they have a synergistic effect. The central mode of action of NSAIDs is not well known, but several studies have shown that in healthy animals (i.e., in the absence of inflammation), NSAIDs increase the threshold to nociceptive simulation. This effect is blocked by co-treatment with naloxone and with atipamezole, a selective alpha-2 receptor antagonist.

Opioids and alpha-2 adrenergic agonists reduce the perception of pain by acting on pain signal projection and the thalamic-cortical structures involved. Benzodiazepines and phenothiazines (acepromazine, Calmivet®), like most general anesthetics, do not have any known analgesic action in animals. *The loss of consciousness of the animal should not be considered as a method of treating pain.*

To overcome pain generated by the surgical procedure, analgesia should be determined by the intensity of pain in each individual case (Fig. 5.10). There are two main classes of analgesics: *NSAIDs* and *morphine derivatives* (opiates).

In rodents, the use of NSAIDs, like aspirin, carprofen, ketoprofen, paracetamol, and flunixin, is often preferred. These compounds have the advantage of not causing sedation or dependence and are very effective against inflammatory pain associated with orthopedic and soft tissue surgery, infection, or trauma. In addition, there are many NSAID preparations, and there are no regulatory restrictions particular to their use or storage. In contrast, morphine and its derivatives such as buprenorphine and butorphanol oblige a much stricter and more regulated use. The choice of analgesic has implications for the larger functions of the body (Table 5.3).

NSAIDs have an impact on immune function and platelet activity and may cause gastrointestinal ulceration. They also have a nephro- and hepatotoxicity that varies from one species to another. NSAIDs are also contraindicated in animals treated with anticoagulants or when coagulation or gastrointestinal functions are the object of the study's experimental protocol.

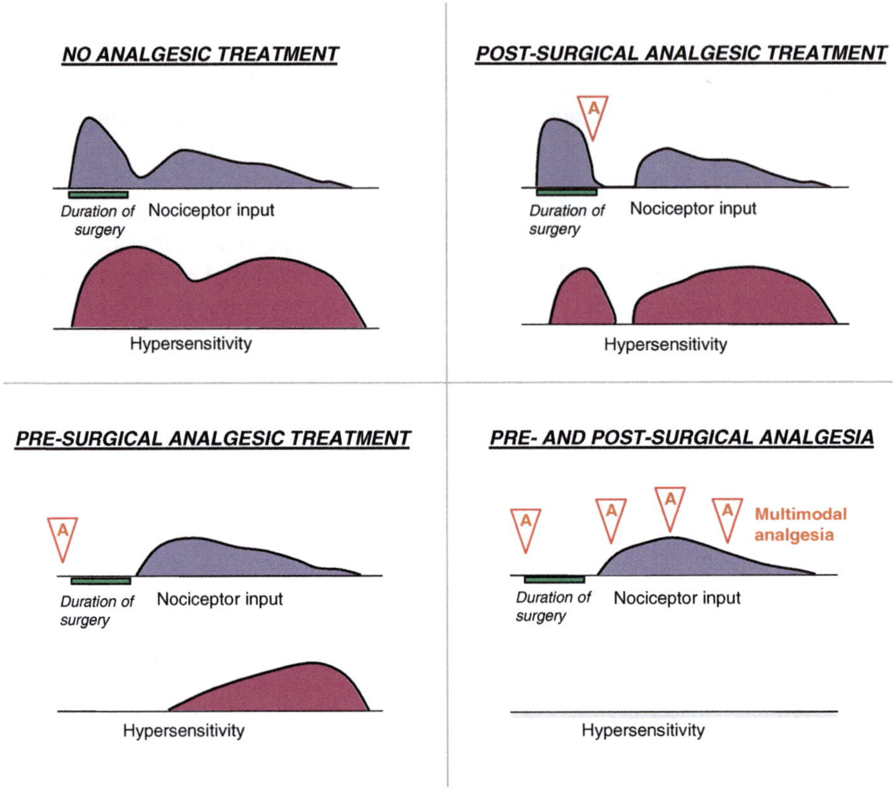

Fig. 5.10 Schematic of preemptive analgesia in preventing sensitization of the nervous system throughout the perioperative period. *A* analgesic drug administration. A typical experience without analgesic intervention is shown in the *top left* diagram, which depicts pain from the initial surgery (nociceptor input, shown in *purple*) and the hypersensitivity (*pink*) that subsequently develops. In the *upper right* diagram, analgesia (A) administered after sensitization may decrease pain somewhat but has little long-term benefit. Analgesia (A) administered before surgery (*lower left*) limits the pain from that stimulus and decreases subsequent hypersensitivity. The most effective preemptive analgesic regimen is initiated before surgery and continued throughout the postoperative period (*lower right* diagram) (Modified from www.aafp.org/afp/2001/0515/p1979.html)

Table 5.3 Principal analgesics used in veterinary surgery of rodents

Molecule	Analgesia	Antipyretic	Action on platelets and bleeding time	Gastrointestinal ulceration, hepato- and nephrotoxicity
Carprofen (Rimadyl®)	+++	+	NO (even chronically)	NO (even chronically)
Paracetamol	+	++	NO	NO
Aspirin	+	++	+++	NO
Flunixin	+++	+	+++	+++
Ketoprofen (Ketofen®)	+++	+	+++	+++

Example of an anesthetic and analgesic protocol for a rat brain implantation procedure (Fig. 5.11).

- Before surgery, the animal receives an analgesic dose of carprofen (Rimadyl®).
- Anesthesia is induced 15 min later by an intraperitoneal injection of xylazine and ketamine.
- Localized subcutaneous injection ("infiltration") of local anesthetics such as lidocaine or bupivacaine affords complete anesthesia at the sites of incision and the pressure points of the stereotaxic apparatus. To minimize bleeding when using lidocaine, it is advisable to use a solution of 5 mg/ml containing 2 % adrenaline (epinephrine) and diluted to ¼ in sterile water. It should not exceed 7 mg/kg, i.e., 40 µl for mouse and 350 µl for rat. The local scarring process is independent of the administration of a local anesthetic (Waite et al. 2010).
- After surgery, animals receive a second dose of Rimadyl®. This treatment should be repeated once a day for at least 2 days after surgery

Management of chronic pain	**Management of acute pain**
Drugs with long duration of action	Perioperative pain
Before the operation: - NSAIDs or related derivatives - Injectable or oral morphine derivatives (buprenorphine) - Transdermal opiate (fentanyl patch)	*Before the operation:* - NSAIDs or related derivatives - Injectable or oral morphine derivatives (buprenorphine) - Transdermal opiate (fentanyl patch)
During the operation: Multimodal analgesia	*During the operation:* Precise and atraumatic surgery
After the operation: - Determine the cause of pain and targets its treatment - Local or regional anesthesia as required	*After the operation:* - Nursing - Infusable analgesics (opiates), and injectable, transdermalor oral drugs (NSAIDs, buprenorphine, fentanyl)

Fig. 5.11 Principal elements in the treatment of chronic or acute pain

Preferred choices are carprofen, particularly effective with a minimal risk of toxicity, and paracetamol, which has no effect on platelet activity and bleeding time, unlike aspirin. Flunixin and ketoprofen are more effective painkillers than aspirin and paracetamol but also have also greater platelet and gastrointestinal toxicities. They are thus reserved for treatment periods of short duration.

The veterinarian can help in establishing the anesthesia and analgesia protocol that suits each type of procedure.

5.3 Anesthesia

General anesthesia induces a temporary and reversible suspension of consciousness, pain sensitivity, and mobility in animals. General anesthetics are typically administered by injection or inhalation (Jensen et al. 2004). Stereotaxic procedures require a general anesthesia in order to obtain a loss of consciousness or artificial sleep, narcosis, a suppression of sensation (especially of pain), suppression of movements, muscle relaxation, as well as preservation of autonomic functions (Harrison 2003).

If the condition of the animal allows the surgery to proceed, the operator should limit stress of the animal prior to anesthesia. For example, a change in the animal's cage environment just before surgery or bringing all of the animals into the surgery room at the same time should be avoided. Stress can be caused by various factors such as the smell of anesthetics and solvents, the movements of the operator, pheromones, and audible or ultrasonic vocalizations of the animal's congeners.

The drugs used for anesthesia can take variable amounts of time before being metabolized and eliminated. Certain of these can persist in the body for up to 1 week after surgery. Thus, special attention should be given in choosing the type of anesthetic, depending upon being the type of experiment (electrophysiology, microdialysis, acute pharmacological treatment, etc.) that will take place after the surgery (Chery et al. submitted; Farber et al. 1997; Hemming 2009; Vizcaychipi et al. 2011).

A good anesthesia involves the combined administration of several molecules in a particular sequence. The different steps in obtaining anesthesia are defined as follows:

- *Premedication* corresponds to the preoperative administration, 30–90 min before the operation, of one or more substances for preventing some of the effects of anesthesia, such as hypersalivation, and to reduce anxiety related to the procedure. This premedication can also allow a reduction in the dose of anesthetic by lowering the basal metabolic activity of the animal.
- *Induction* of anesthesia is the most delicate part of anesthesia because of the complications that may arise, such as agitation, glottal spasms, hypertension, tachycardia or bradycardia, respiratory depression, apnea, atrial fibrillation, or cardiac arrest. Induction occurs immediately following administration of the anesthetic and should be as short as possible.
- *Anesthesia* is maintained in a continuous manner to keep the animal at a surgical level throughout the procedure. Repeated reinjections or continuous infusion of anesthetic can be used to this end.
- *Awakening* from anesthesia begins after stopping the administration of anesthetic and may be affected by the consequences of the surgical procedure.
- *Recovery* of brain functions (motor, sensory, and cognitive) and normothermia signal the return to a normal physiological state, while the elimination of certain anesthetics may continue over several days.

The different steps in the anesthetic procedure (premedication, induction, maintenance, etc.) should not be confused with the different stages of anesthesia that define its level (depth)

5.3.1 The Different Stages of Anesthesia

The stages described below may vary in duration and intensity, depending on the anesthetic used. In addition, each species has a particular behavior during anesthesia, sometimes with marked individual variations. As anesthesia deepens, reflexes disappear in the following order:

- *Palpebral* or *blink reflex:* lightly touching the eyelids causes the animal to blink. In this case, the animal is not sufficiently anesthetized.
- *Pinch reflex*: pinching the toe or footpad of the animal with the fingers causes discomfort. If the animal withdraws its paw, it is not adequately anesthetized. Otherwise, if sufficiently anesthetized, the animal should feel no pinch or pain.

Definition of the Different Stages of Anesthesia

- *Stage 1 or induction:* evidenced by voluntary excitation with increased cardiorespiratory rhythm.
- *Stage 2 or stage of delirium and involuntary excitation*: seen as a loss of progressive awareness and exaggerated responses to normal stimuli. The animal may have periodic apnea, more or less marked, and vomiting, hence the importance of fasting the animals overnight to avoid any risk of aspiration and choking. This stage should be kept as short as possible to minimize the occurrence of accidental injury to the animal.
- *Stage 3 or stage of general anesthesia* consists of four phases, including surgical anesthesia.
  - *Phase 1 – light anesthesia*: most reflexes (palpebral, pinch, and corneal) are reduced but still present.
  - *Phase 2 – medium anesthesia:* most reflexes (blink, pinch, corneal) are absent and the muscles are relaxed. Surgeries are performed at this level; this is the stage of surgical anesthesia.
  - *Phase 3 – deep anesthesia*: pupillary reflexes (change in the diameter of the pupil in response to the intensity of light that falls on the retina of the eye) are diminished or absent. A decrease in muscle tone affects the intercostal muscles and spontaneous breathing, thus requiring assistance (artificial respiration).

  - *Phase 4 – excessive anesthesia*: complete absence of corneal reflex and paralysis of muscles, including respiratory muscles (intercostals and diaphragm).
- *Stage 4 – irreversible anesthesia*: characterized by respiratory arrest, followed by cardiovascular collapse and death within 1–5 min.

- *Corneal reflex:* carefully touching the cornea of the eye with a tuft of cotton causes the eyelids to blink. Once the corneal reflex disappears, the animal is under excessive anesthesia.

5.3.2 Monitoring Vital Signs

Anesthesia can induce mortality and morbidity by altering the metabolism of the animal. Given their small size, rodents are quite susceptible to hypothermia and dehydration, especially during inhalational anesthesia. The following section aims to define and standardize a rigorous procedure for monitoring the experimental animal before, during, and after surgical interventions. Emphasis will be given to the principal points to consider and will not detail the multiplicity of methods and techniques that can be employed. While the priority is to monitor the proper *depth of anesthesia* during the operation by checking the various reflexes, attention also needs to be given to maintaining general homeostasis, body temperature, and proper hydration of the animal.

The loss of consciousness of the animal requires that the operator ensure its *physical protection and integrity*: protect the cornea if the eyes are open, prevent against any discomfort or difficulty in breathing, and avoid any prolonged pressure on tissues. During prolonged anesthesia, or in the case of a more fragile newborn subject, or a transgenic animal with an altered metabolism, the maintenance of physiological stability of each subject needs to be considered and assured.

General anesthesia induces a depression of thermoregulatory centers, a lowering of metabolic activity and a peripheral vasodilation that can lead in a matter of minutes to a drop in body temperature and a fatal hypothermia under 32 °C. Hypothermia during an operation can lead to difficulties in awakening, discomfort, and complications from poor blood perfusion, inadvertent coagulation, and ischemia. Use of a heating pad placed under the abdomen of the animal during the operation and space heating during the waking phase is a necessity. Some models of heating pads or blankets are equipped with a thermostatically controlled fine regulation of internal body temperature via a rectal probe (Fig. 5.12). This type of device should be used with an insulating layer placed between the heating pad and the cold surfaces of the stereotaxic base, to avoid heat loss to the metal.

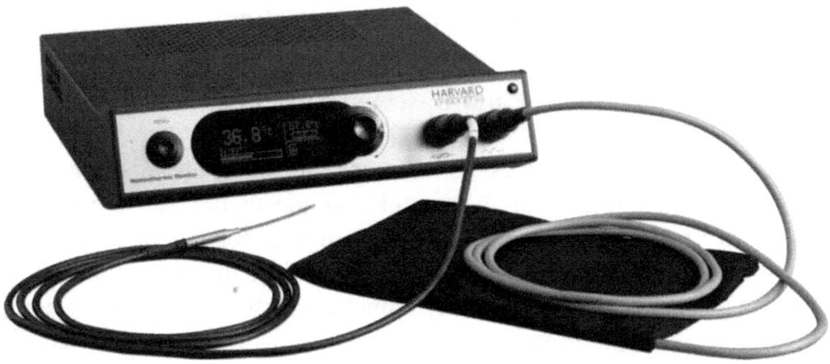

Fig. 5.12 Heating blanket and rectal probe for a thermostated temperature control of 37 °C (Harvard Apparatus®)

Anesthesia can provoke cardiorespiratory depression of more or less prolonged duration. Thus, two possible risks arise: poor oxygenation and a drop in blood pressure, both sources of neurological complications and tissue damage. There are several devices for monitoring blood gases (oximetry, capnography) and arterial pressure in humans and domestic animals. These also exist for rodents although they are not widely considered as being necessary for operatory procedures of relatively short duration. They are, however, essential for reducing morbidity in the case of long, complex surgeries (Drummond 2000). Even without specialized equipment, it is still possible to ensure that there is no difficulty in breathing, that the respiratory rate is sufficient, and that the color of the mucous membranes is satisfactory during the operation. With an inhalational anesthesia, it is recommended that a supplement of oxygen be used during surgery to avoid hypoxia (Fornari et al. 2012) and that ambient air is not used. In addition, in such procedures, it is essential to prevent bronchial hypersecretion, by injecting muscarinic cholinergic receptor antagonists such as atropine or glycopyrrolate (Robinul®).

In the event of apnea or hypoventilation, the first action is to reduce the depth of anesthesia and assist in the recovery of an appropriate respiration by manual massage of the chest and oxygen supplementation. Sublingual or intravenous administration of doxapram (Dopram®) (1–5 mg/kg) is highly effective for treating apnea and respiratory arrest. A lengthy anesthesia and/or intraoperative blood loss can lead to dehydration and hypovolemia. Electrolyte disorders accompanied by acidosis due to glucose consumption can be offset by subcutaneous injection of physiological or enriched solutions (e.g., Ringer lactate®) containing glucose or other macromolecules. Moreover, the duration and depth of anesthesia should be limited to the strict minimum necessary for the proposed procedure. The response of an animal to anesthesia depends on individual characteristics and several factors including age and weight. Appropriate doses of anesthesia should be respected and vigilance maintained throughout the duration of the operation (Appendix 8) (Richardson and Flecknell 2005). When permitted, the use of agents that reverse anesthesia can induce a rapid awakening of the animal.

5.3.3 *Different Types of General Anesthetics*

5.3.3.1 Inhalational (Volatile and Gaseous) Anesthesia

In terms of safety and effectiveness, inhalational anesthesia is superior to most parenteral anesthetics (Sonner et al. 2003). Inhaled anesthetics are removed from the blood primarily by expiration and are therefore less dependent on the metabolic breakdown of the molecule per se. Toxicity to the organism is thus reduced. The depth of anesthesia is easily and rapidly adjustable, which is useful for compensating the variability of the effective dose between individual animals and from one species to another. The complexity and cost of the equipment required for administration pose an obstacle to their use. In addition, volatile liquid products need to be stored in an isolated room or cabinet since vapors are flammable and toxic to personnel over regular and extended exposures. For small species, induction under a bell jar is usually followed by maintenance using an anesthetic mask. Inhalational anesthesia requires specific precautions: appropriate premises and specialized equipment suitable for its administration, cartridges of activated charcoal for scavenging excess compound in the effluent circuit, regular controls and maintenance of this material, and limiting the exposure time to personnel.

The minimum alveolar concentration, or MAC (Sonner et al. 2003), is the percentage of inhaled anesthetic necessary and sufficient for anesthetizing and blocking any reflex reaction to a standardized noxious stimulus in 50 % of subjects. Usually this percentage is multiplied by a factor of 2 or 3 for induction and by a factor of 1.2 for the maintenance of anesthesia. The duration of induction varies with the alveolar concentration and the effective circulating blood level. These factors likewise determine the recovery time once the administration of anesthetic is stopped. To have a rapid induction, which is ideal, the anesthetic agent should be easily absorbed into the blood. However, an excessive absorption or accumulation in the blood compartment poses a risk for overdose and a longer waking time. Since all of these agents are partially metabolized by the body, their impact on the individual may be amplified if there is any preexisting organ dysfunction (e.g., hepatic).

Most volatile anesthetics are halogenated compounds, such as isoflurane, sevoflurane, desflurane, enflurane, and halothane (Jurd et al. 2003). Isoflurane, the most commonly and currently used (Messier et al. 1999), has a rapid induction and awakening due to its low solubility (gas/blood partition coefficient of 1.4). It provokes little or no salivation or tracheobronchial secretions. By adjusting the vaporizer, the level of anesthesia can be rapidly adjusted. Isoflurane, as well as the other anesthetics of this class, produces a decrease in blood pressure and dose-dependent neuromuscular depression. It is minimally metabolized (<0.2 %) and almost entirely eliminated unchanged via the lungs. Recommended levels for use in the rodent are 2–3 % for induction and 0.25–2 % for maintenance. If the apparatus used for anesthesia permits, nitrous oxide (N_2O) can be added to the mixture (50–70 %) to reduce the concentration of isoflurane used. However, its use for short surgical procedures in the rodent remains debatable. As isoflurane has little or no analgesic properties, an analgesic should be associated with its use.

The use of halothane has fallen out of use because of its hepato- and nephrotoxic properties in humans during intense and prolonged exposures. It induces cardiovascular depression and a rare but fatal postoperative malignant hyperthermia. However, it is preferred over a parenteral anesthetic for long procedures in rodents, when the equipment is available. The volumes used are extremely low, which limits the exposure of the operator, particularly when working in an environment where the renewal of air is 15–20 volumes per hour.

5.3.3.2 Parenteral Anesthesia

The main drugs used in parenteral anesthesia in animals are presented here. Their combination allows synergistic actions and a significant lowering of the doses used. Certain agents such as opiates come under specific regulations of storage and use. In general, injectable anesthetics are metabolized by the liver and excreted by the kidneys. For debilitated patients or those with liver and/or kidney disease, an inhalational anesthetic is preferable since the majority of its elimination is via the respiratory system and has minimal impact on hepatorenal functions.

Thiazines

Xylazine hydrochloride (Rompun®) and medetomidine (Domitor®) are similar products that act as alpha-2 adrenergic receptor agonists. They produce CNS depression resulting in sedation, moderate analgesia, and secondary hypotension and bradycardia. Xylazine also produces respiratory and cardiovascular depression, as well as hypothermia. Medetomidine is preferred, since it has a rapid onset of action and more powerful anesthetic effects. In experimental procedures, these compounds are used in combination with ketamine (Imalgene®) for anesthesia of rodents. Their effects can be reversed with alpha-2 receptor antagonists: yohimbine for xylazine (Rompun®) and atipamezole (Antisedan®) for medetomidine (Domitor®). Atipamezole and medetomidine have been specifically formulated for veterinary use, allowing a shorter waking phase and a significantly safer use of the anesthetics.

Dissociative Anesthetics

Dissociative anesthetics such as ketamine (Imalgene®) are easy to use and have a wide therapeutic index[3] in most laboratory species (Stokes et al. 2009). These

---

[3] The dose of anesthetic that induces lethality of 50 % of a sample of animals is expressed as the LD50 value. This value gives an idea of the amount that can be safely injected in an animal. For ketamine, the LD50 is 60 mg/kg by the intravenous (IV) route and 600 mg/kg by the intramuscular (IM) route. The effective doses for anesthesia are 4.5 mg/kg IV and 6.5–13 mg/kg IM.

agents have limited analgesic activity, particularly against visceral pain, and must be used in combination with an analgesic. Ketamine induces a state of catalepsy characterized by a suspension of involuntary movement, with respiratory depression and limited myorelaxation, but no impairment of cardiovascular function. Ketamine is used in combination with other molecules that have a strong muscle relaxant property, such as xylazine, and with painkillers such as Butorphanol® or Fentanyl® (*mu*-opioid receptor agonists) to compensate for its weak analgesic efficacy. At the level of the CNS, ketamine depresses thalamocortical activity but stimulates the limbic and reticular formation, producing dissociation. Ketamine increases cerebral blood flow and can also raise intracranial cerebrospinal fluid pressure.

Besides an action on glutamate receptors, ketamine (Imalgene®) has sympathomimetic effects that reinforce the actions of adrenaline and noradrenaline. Simultaneously, at the periphery, it has a parasympathomimetic action manifested by abundant salivary, bronchial, and lachrymal secretion that can be blocked using atropine and glycopyrrolate. Ketamine is a highly acidic solution, injectable by intramuscular, intravenous, or intraperitoneal routes. Intramuscularly, the induction time is 3–5 min, with about 20 min to reach a surgical level of anesthesia in most laboratory species. In rats and mice, intraperitoneal administration is more painful and increases both the induction and awakening times.

It is generally better to use the intravenous route of injection even for small animals since the necessary volumes of administration are very low and the time of onset is rapid. A surgical level of anesthesia can be attained within ten minutes and, if necessary, with follow-up injections of one-half and one-quarter of the initial dose. Dissociative anesthetics used alone generally preserve the swallowing reflex, which reduces the risk of choking and aspiration in case of vomiting. If used as an inducing anesthetic before switching to an inhalational one, it is preferable to apply a local anesthetic spray or gel before intubation. Ketamine causes acclimation and an increase in intraocular pressure, corneal and pupillary reflexes are intact, and the eyes remain open. It is thus recommended to apply an ophthalmic cream to prevent the eyes from drying out.

Other Anesthetics

- *Propofol* (Diprivan®) is a hypnotic sedative with low analgesic efficacy. It can be used for the induction or maintenance of general anesthesia and should be used in association with an analgesic. Propofol provides very fast induction and awakening and can be injected continuously for prolonged periods without increasing the awakening time (Jurd et al. 2003; Nelson et al. 2002). The product is available as a ready to use emulsion, strictly for intravenous injection only. It is relatively expensive and once opened the ampoule should be used within 48 h. Although safe, it can still cause apnea, bradycardia, and hypotension in cases of overdose.
- *Urethane* is a long-acting anesthetic (8–10 h) that produces minimal cardiopulmonary depression. The drug can be used for lengthy procedures in rodents. It is

carcinogenic[4] (Mason et al. 2000; Li et al. 2013) and has *no analgesic properties. Its use should be limited to acute terminal procedures (i.e., without recovery of the animal), with detailed justification for excluding any other choice of anesthetic.*

- *Tribromoethanol* (Avertin®) is a fast-acting anesthetic used for rodent surgery at doses of 225–240 mg/kg intraperitoneally or 80–120 mg/kg intravenously. It causes a rapid induction, 15 min of surgical anesthesia, and requires up to 90 min for the recovery of the animal. Tribromoethanol is not commercially available and must be prepared as needed. Sterile preparatory procedures are essential, and the solution should be stored in the dark at 4 °C to prevent its degradation. Initially acclaimed by the IACUC (Institutional Animal Care and Use Committee) the use of tribromoethanol remains subject to controversy. The inappropriate conservation and instability of prepared solutions can result in intense inflammation of the peritoneum and visceral pain that has not been adequately evaluated. The Swiss Federal Veterinary Office no longer accepts its use following observations of abdominal adhesions and necrosis upon autopsy. French regulations limit its use since there are many veterinary anesthetics that can be used in its place.

5.3.3.3 Examples of Commonly Used Associations of Parenteral Anesthetics

By combining an anesthetic and an analgesic, the effective dose of each can be reduced, sometimes dramatically, thus ensuring a good quality of anesthesia for the patient. It is recommended to administer the agents separately and not to mix them in the syringe unless one is certain that the compounds are chemically compatible and that no micro precipitates are formed. In any event, mixed or diluted compounds are prepared only as required and should never be kept for future use.

- *Ketamine/xylazine*: xylazine hydrochloride (Rompun®) associated with ketamine (Imalgene®) provides a good quality of surgical anesthesia, with short induction and good myorelaxation. Surgical anesthesia is easily maintained by reinjections of a half and quarter dose after an initial phase of 30–45 min. A mixture of 3/4 part ketamine and ¼ part xylazine is administered by intramuscular, intraperitoneal, or intravenous injection. This mixture causes more cardiorespiratory depression than ketamine alone but remains indicated for painful procedures.
  As previously mentioned, yohimbine (an alpha-2 adrenergic receptor antagonist), injected at a dose of 0.1–0.15 mg/kg by intramuscular or intravenous routes, reduces the time of postanesthetic recovery.
- *Ketamine/medetomidine* (Fig. 5.13): medetomidine (Domitor®), when used in association with ketamine (Imalgene®), provides a good quality of surgical anesthesia with a short induction time (less than 5 min), good muscle relaxation, and easy maintenance by reinjection of half and quarter doses after an initial phase of 30–45 min (Bekker and Sturaitis 2005).

---

[4]Report on Carcinogens, Twelfth Edition (2011), US Department of Health and Human Services.

With this combination, a subcutaneous or intramuscular injection of atipamezole (Antisedan®) should not be given before a minimum of 30–40 min following induction. Waking time is faster and this limits the occurrence of seizures associated with ketamine alone (Imalgene®) which is then largely eliminated by metabolism. Note that atipamezole (alpha-2 adrenergic antagonist), used at 0.25 mg/kg intravenously or 0.2 mg/kg intramuscularly, is a specific antidote for medetomidine and should not be used in any other indication. In severe respiratory depression, atipamezole (Antisedan®) can be injected intravenously to reverse the action of medetomidine (Domitor®) and facilitate the resuscitation of a valuable experimental animal. Management of analgesia should then be reassessed. However, outside of an emergency, a more gradual reversal by subcutaneous injection of atipamezole is preferred.

The investigator should make a record of all compounds administered to the animal (anesthetic, analgesic, and other), indicating the doses, times, and routes of administration, and monitor any potential problems during the intraoperative and recovery phases (Appendices 2A and 2B). These reports should be available for regulatory inspections and will be especially useful for personnel performing the postoperative care.

Practical examples for rats	**Practical examples for mice**
Ketamine 90–100 mg/kg + xylazine 10 mg/kg ***by intraperitoneal route***	Ketamine 100 mg/kg + xylazine10 mg/kg ***by intraperitoneal route***
Ketamine 75 mg/kg + medetomidine 0.5 mg/kg, with a dose of atipamezole of 0.2–0.5 mg/kg ***by intraperitoneal route***	Ketamine 75mg/kg + medetomidine 0.5 mg/kg, with a dose of atipamezole of 0.2–0.5 mg/kg ***by intraperitoneal route***

Fig. 5.13 Examples of anesthetics used for conventional surgical procedures in rodents

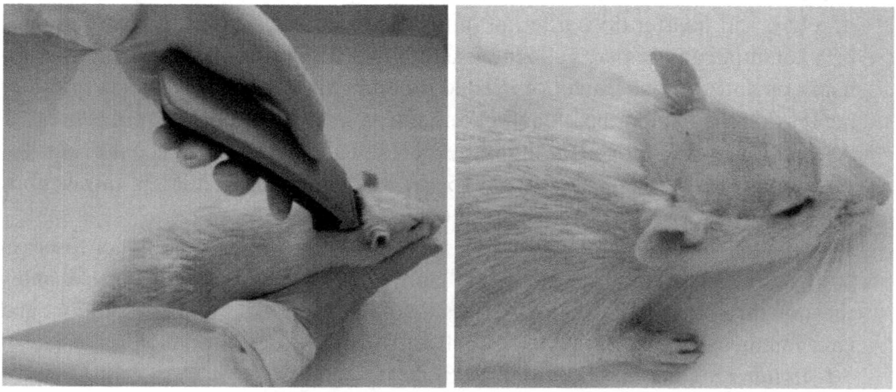

Fig. 5.14 Using the hair clipper to shear a sedated animal just prior to surgery

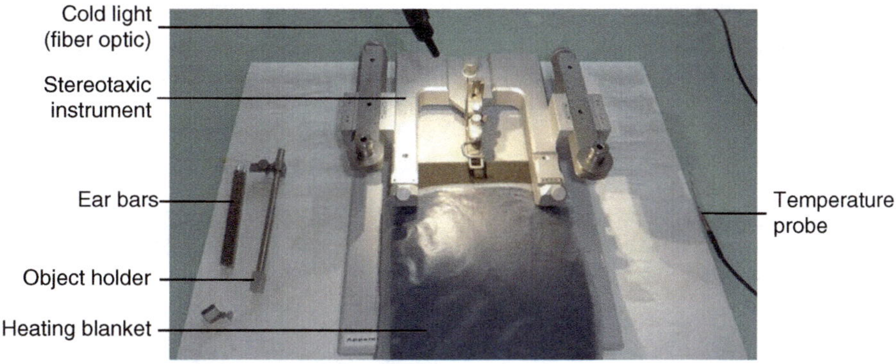

Fig. 5.15 Stereotaxic instrument equipped with a thermostatically controlled heating blanket and temperature (rectal) probe. Also shown in the photo is the object (e.g., electrode or cannula) holder and the ear bars

5.4 Shearing

Shearing is an important preparatory step because it eliminates hair and facilitates disinfection of the skin. Shearing should be as close as possible without injuring the skin, since any inflammation tends to increase the risk of local superinfection and weakens the tissues. It must be performed on a sedated or anesthetized animal (Fig. 5.14) just prior to surgery in the preparation zone, so as not contaminate the surgical area.

5.5 Setting Up the Animal in the Stereotaxic Instrument

Placing the animal in the stereotaxic apparatus can begin once a level of surgical anesthesia, characterized by a loss of reflexes and a regular rhythm of spontaneous breathing, is reached. When all of the equipment is ready and available on the table in the proper aseptic conditions, the animal is placed in the instrument on a heating blanket or pad (Fig. 5.15) as described in the following steps.

5.5.1 Placement of the First Ear Bar

To begin, one of the two bars must first be securely fixed in the instrument (the left bar is shown in Fig. 5.16). The head of the animal is gently grasped by the sides to guide the ear canal towards the fixed bar. The head is then carefully maneuvered horizontally onto the bar so that the tip of the bar is positioned behind the aural bone spur, as shown by the arrow in Fig. 5.17. The correct placement is evidenced by a soft characteristic click, not to be confused with a louder snap that indicates the rupture of the eardrum if the bar has been inserted too far.

Note: Do not hold the head of the animal by placing a finger around its neck, which could block the trachea. The animal is anesthetized, and a poor maintenance of the position of the head can cause asphyxiation.

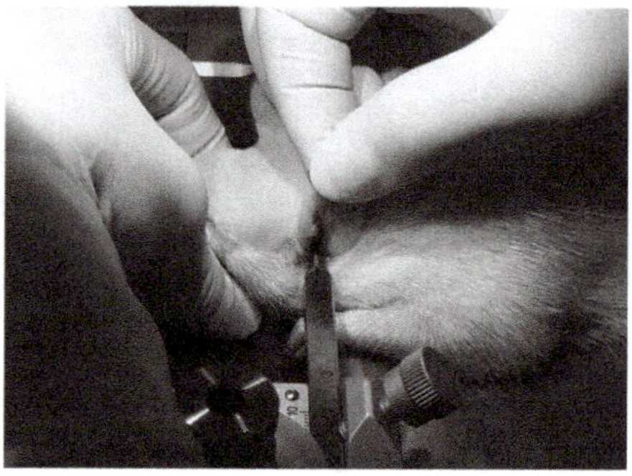

Fig. 5.16 Photo showing the gesture made in placing the ear canal of the animal in line with the first ear bar

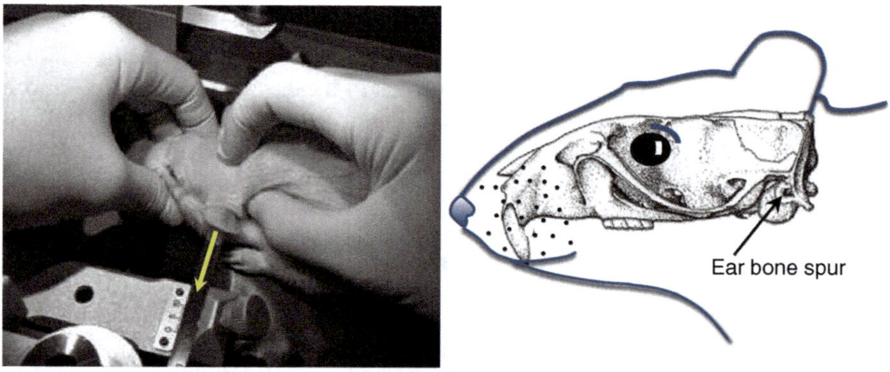

Fig. 5.17 Placing the ear bar just behind the ear bone spur

5.5.2 Placement of the Second Ear Bar

The head of the animal is maintained in a horizontal position to guide the second bar into the other ear canal (Fig. 5.18). If the ear bars have been properly placed, the head of the animal cannot be moved from side to side. Before proceeding, the operator should check that the only movements possible at this point are an antero- and retroflexion.

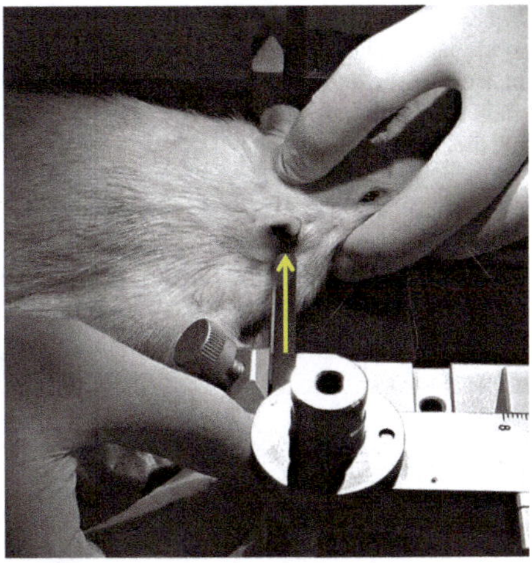

Fig. 5.18 Placement of the second ear bar

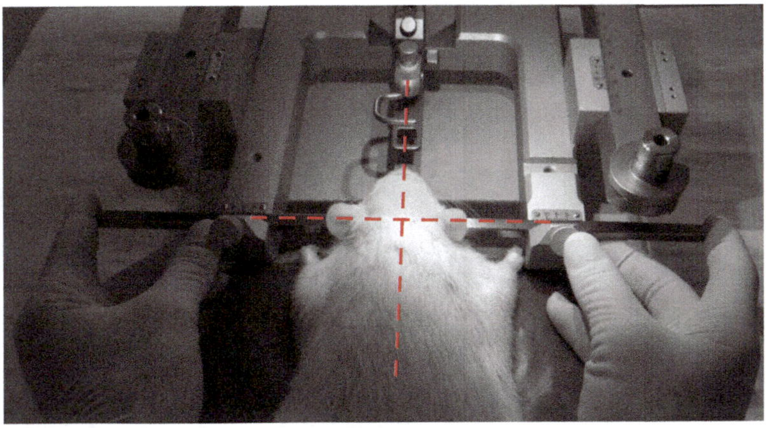

Fig. 5.19 Aligning the animal's head with the tooth bar

5.5.3 Placement of the Tooth Bar

When the ear bars are properly in place, the head of the animal is aligned with the centerline of the stereotaxic frame (Fig. 5.19).

Once the head is properly aligned, the upper incisors of the animal are fixed on the tooth bar. To do this, the bar is moved towards the animal, the mouth is opened, and the tooth bar is slid behind the upper incisors (Fig. 5.20a). By gently pulling the head of the animal forward, the bar moves into the naturally correct position behind the teeth (Fig. 5.20b).

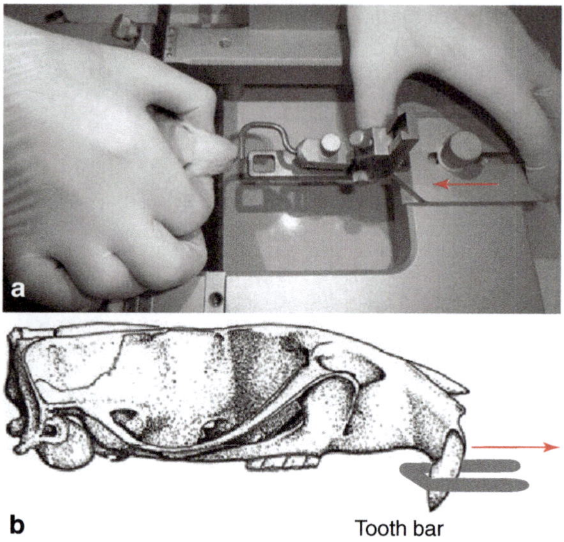

Fig. 5.20 Placement of the tooth bar behind the upper incisors of the animal. (**a**) Top view, (**b**) lateral view of the position of the incisor bar

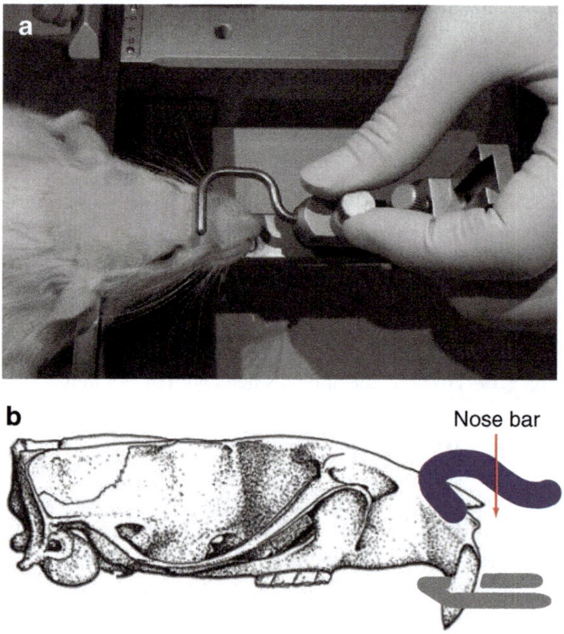

Fig. 5.21 Fixing the nose bar. (**a**) Top view, (**b**) lateral view of the position of the nose bar

5.5.4 *Fixing the Nose Bar*

Finally, the nose bar is lowered to lock the snout in position (Fig. 5.21a, b). Fixation of the nose bar with the setscrew must be sufficiently firm, but without applying excessive pressure, so that the head is immobilized without injuring the animal. When the incisor bar and nose are properly placed, no more head movement is possible.

5.6 Preparation of the Operating Zone

Before proceeding with the disinfection of the surgical site, a rectal temperature probe, if available, should be put in place at this time. Vaseline can be used to minimize the trauma and discomfort of insertion. After placing the animal on the ear bars under sterile conditions, the skin where the incision will be made is disinfected by washing with sterile gauze soaked with antiseptic soap, alcohol, and an antiseptic solution. The principle of the washing procedure is to progress centrifugally, away from the incision site towards the periphery without coming back. If the edge of the shaved zone is touched, the gauze pad should be discarded. Antiseptics disinfect the skin and eliminate the local microorganic flora. The spectrum of antiseptic activity varies by product and by the quality of its application. A suitable product should be chosen and its duration of contact should be strictly respected.

Note: Most concentrated antiseptic solutions should not be used on internal tissue or mucosa. It is advisable to use the same antiseptic under various forms: chlorhexidine soap and alcohol solution, or soap solution and aqueous iodine solution. The principal broad-spectrum antiseptics used for disinfecting an operating zone are listed in Table 5.4. Currently, the two most widely used skin antiseptics are chlorhexidine and povidone-iodine (Betadine®).

Note: The use of certain colored products may subsequently interfere with visual inspection of the tissues, masking points of hemorrhage, or areas of irritation or ischemia.

Table 5.4 The principal antiseptics used in veterinary surgery and their spectrum of action

			Gram + bacteria	Gram – bacteria	Naked virus	Enveloped virus	Fungus	Spores	
Membrane denaturant	Quaternary ammonium	Benzalkonium chloride	+	±	±	+	±	–	Dermacool® Mercryl® Not for use on mucous membranes; retards healing
	Bisbiguanide	Chlorhexidine	+	+	±	+	±	–	Hibitane® Mercryl® Not for use on mucous membranes; retards healing
	Diamidine	Hexetidine	+	–	–	–	–	–	Duphaderm®
	Alcohol	Ethyl alcohol/ethanol	+	+	+	+	+	–	Irritation and pain. Not for use on mucous membranes, wounds, corneal or connective tissue
Oxidizing agent	Iodinated compounds	Alcoholic iodine solution	+	+	+	+	+	+	Tincture of iodine, iodized alcohol, iodized glycerin No occlusive dressing, not for use on burns
		PVPI– Polyvinylpyrrolidone iodine	+	+	+	+	+	+	Vétédine®Bétadine® Irritant, no occlusive dressing, not for use on burns
	Hypochlorite	Sodium hypochlorite	+	+	+	+	+	+	Dakin® Unstable overtime
	Peroxide peroxy salt	Hydrogen peroxide	+	+	+	±	±	–	Hemostatic (for capillary hemorrhage). Irritant at high doses, not for use with potassium permanganate

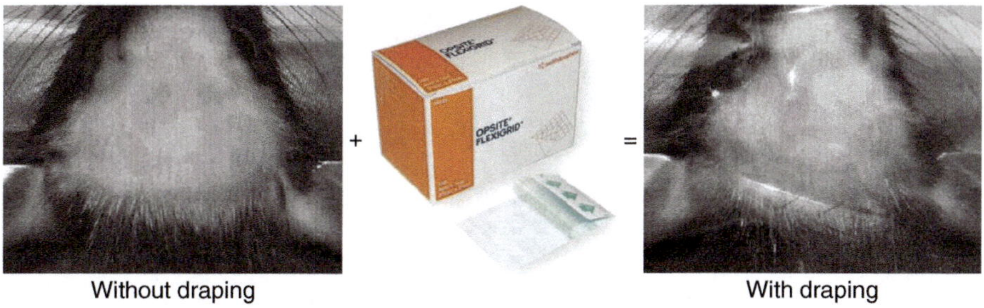

Fig. 5.22 Head of the animal before (*left*) and after draping (*right*)

5.7 Draping

The role of surgical drapes is to maintain sterility of the cleaned zone and to prevent any contamination from adjacent untreated areas. A wide variety of sterilizable tissue fields, reusable or disposable, are commercially available. For surgery of the rat or mouse nervous system, a disposable self-adhesive transparent field, such as Opsite Flexigrid® and Smith and Nephew®, permits visual monitoring of the vital functions of the animal (Fig. 5.22). The drape should cover the entire animal and can eventually be completed by one or more drapes to the adjacent areas to extend the surface of the sterile zone. It is important to ensure an overall coherency in the draping procedure. For a better visual appreciation of the surgical procedures described in this manual, photographs in the illustrations were taken without the surgical drapes in place.

5.8 Incision and Exposure of the Cranial Surface

A scalpel with a new blade is used to incise the scalp. The gesture must be confident and precise to avoid hacking the edges of the wound that complicates healing. The linear incision starts rostrally, slightly in front of eye level, continues caudally to about 0.5 cm behind the ear bar alignment line and involves all layers of the skin in its entirety (Fig. 5.23).

5.9 Stereotaxic Landmarks

The different suture lines of the four bone plates of the skull surface must be visible to correctly identify the stereotaxic landmarks, *bregma* and *lambda* (Fig. 5.24). The quality of the approach to the target structure depends on the precise determination of these two reference points. Bregma corresponds to the intersection of the sagittal and frontal sutures at the rostral level of the animal's head. Lambda, whose shape is

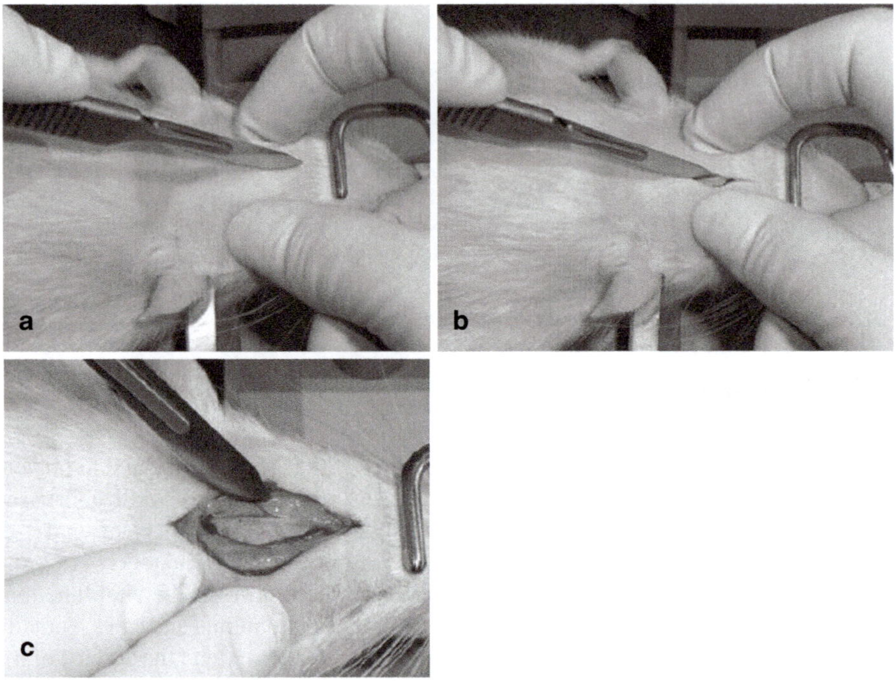

Fig. 5.23 Different steps of the skin incision. (**a**) Puncture of the skin with the tip of the scalpel blade at 90° to begin the incision. (**b**) Slicing with the edge of the blade positioned at 45° holding the scalpel like a violin bow, following through while applying pressure with the index finger on the blade. (**c**) End of the incision by returning the blade to 90°. The connective and periosteum are then retracted laterally on either side of the midline, using a spatula or a rasp to expose the bone

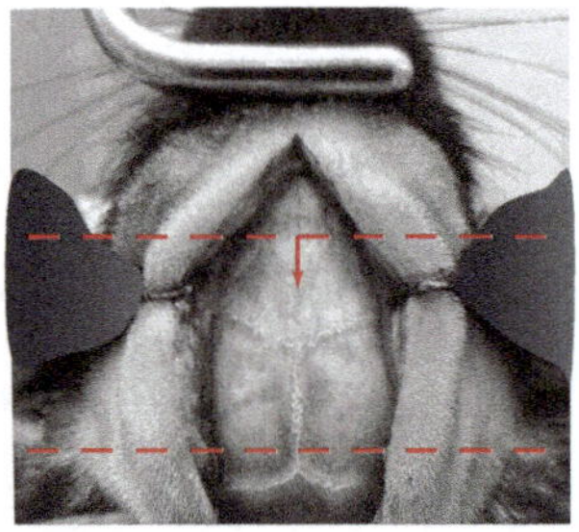

Fig. 5.24 Photograph of the operated zone and skull surface after retraction of the overlying connective tissue and periosteum in a Long-Evans rat. The *dotted red lines* represent the axis of the alignment of the eyes (*upper line*) and ear bars (*lower line*). The *arrow* indicates the direction of the incision

reminiscent of the Greek letter *lambda*, is a virtual point located on the midline and at the base of the triangle formed by the lambdoid sutures and the sagittal suture (Fig. 5.25). Generally, lambda is the point directly in line with the virtual interaural line represented by the axis of the ear bars (Figs. 5.25, 5.29, and 5.30). It may sometimes be situated just anterior to the interaural line (Fig. 5.26, according to Paxinos and Watson 2004) but never posterior. As a consequence, the distance between bregma and lambda points can vary from 8.7 ± 0.3 mm.

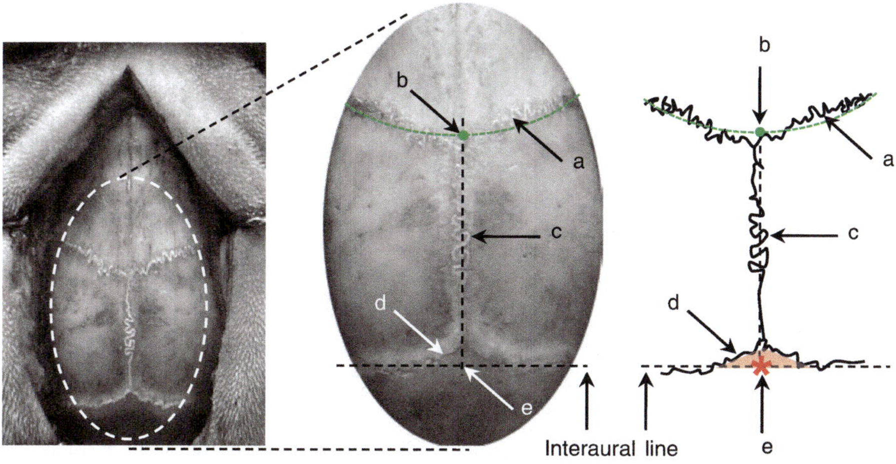

Fig. 5.25 The different bone plates of the rodent skull and their lines of approximation (*suture lines*). In this example, lambda (*red asterisk*) is situated on the interaural line. The triangle formed by the lambdoidal and sagittal suture lines is shown in *light red. a* Frontal or coronal suture. *b* Bregma. *c* Sagittal suture. *d* Lambdoidal or occipital suture. *e* Lambda located on the interaural line

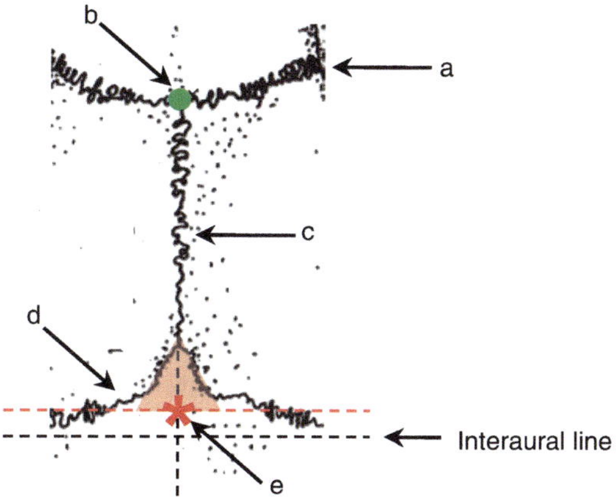

Fig. 5.26 Diagram showing the different bone suture lines on the skull surface of a Wistar rat (From Paxinos and Watson 2004). In this example, lambda (*red asterisk*) is located rostral to the interaural line. The triangle formed by the lambdoidal and sagittal suture lines is shown in *light red. a* Frontal or coronal suture. *b* Bregma. *c* Sagittal suture. *d* Lambdoidal or occipital suture. *e* Lambda, located on the interaural line

The point corresponding to the intersection of the sagittal and lambdoid (or lambdoidal) sutures is sometimes called lambdoid. *Note:* this point may vary from one animal to another and from one species to another.

To ensure the accuracy of the coordinates of lambda, one must make certain that the distance between this point and the bregma point is equal to about 8.7 ± 0.3 mm (Fig. 5.27). If the value is greater than 9 mm or less than 8.4 mm, the determination of one or both of these two points is incorrect.

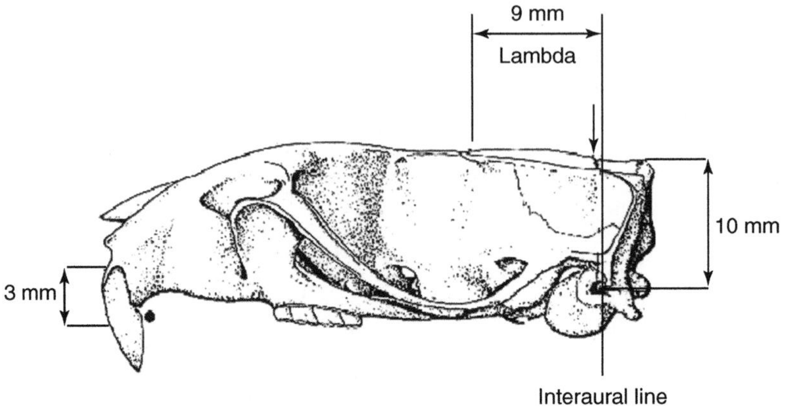

Fig. 5.27 Side view of the bone sutures on the skull surface of a Wistar rat (From Paxinos and Watson 2004) showing the lambda point (*arrow*) in relation to the interaural line

5.10 Surveying and Taking Coordinates

Once the bregma and lambda reference points have been well identified, the operator can proceed with taking their anteroposterior (AP), mediolateral (ML), and dorsoventral (DV) coordinates. To do this, the tip of the instrument chosen to attain the target structure (e.g., microinjection needle, cannula guide, microdialysis probe, electrode) is used. In the case of a microinjection needle fixed to the object carrier, the needle is moved by means of the anteroposterior and mediolateral micromanipulators so as to align its point on bregma and lambda (the order is irrelevant). For example, after placing the needle onto the surface of the bregma point, the AP, ML, and DV coordinates are recorded. The object carrier is then raised on the dorsoventral axis and moved along the anteroposterior axis to the lambda point. When the needle is placed directly above lambda, the object holder is carefully lowered so that the tip of the needle is flush with the reference point. The AP, ML, and DV coordinates lambda are then noted. To validate the horizontality of the skull, the ML and DV coordinates of the two reference points should be identical, with a tolerance of ± 10 μm. The coordinates of each point are recorded in a laboratory notebook (see Appendix 2A). A binocular loupe or magnifying headset (Fig. 5.28) is sometimes necessary to carry out this operation.

5.10.1 *When the Anatomy Makes It Impossible to Identify Bregma and Lambda*

Sometimes it is not possible to identify lambda and bregma using the visual anatomical landmarks. In this case, a geometric construct is used to define these points for determining their coordinates (see Figs. 5.29, 5.30, and 5.12; Blasiak et al. 2010). Having identified the two lateral bone edges, the distance between them is measured. The midpoint of segment "c" is determined. The mediator

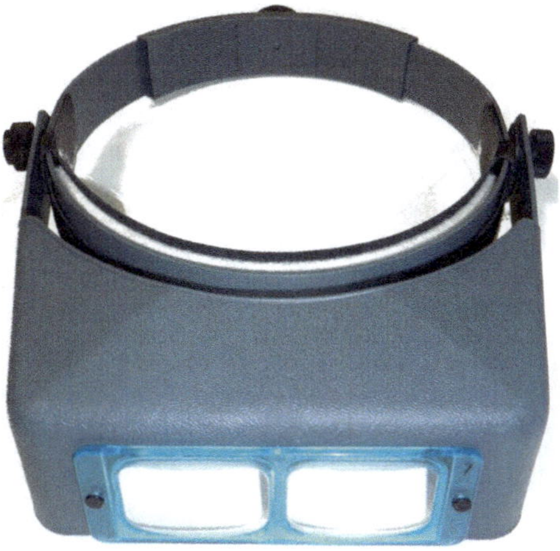

Fig. 5.28 Magnifying headset commonly used in rodent neurosurgery

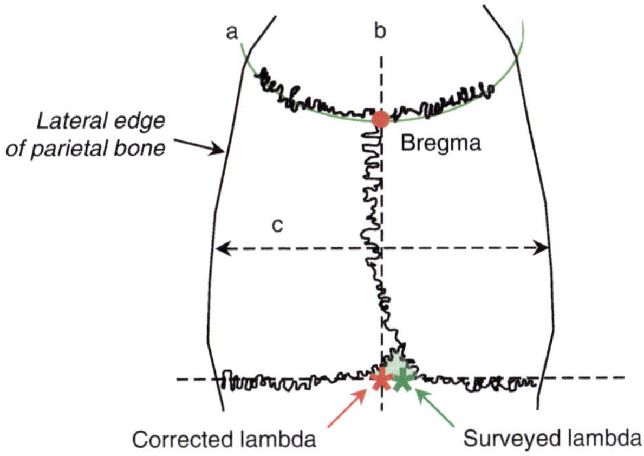

Fig. 5.29 Diagram showing the skull bone sutures in an animal having an asymmetric junction of bone plates. In this example, lambda is not aligned with bregma. The curve "a" designates the frontal sutures. The midline "b" is equidistant from the two lateral bone edges. This distance is defined by the segment "c"

(perpendicular bisector) of this segment, line "b," intersects the curve "a" that forms the frontal suture at a point that is defined as bregma. Similarly, line "b" bisects the interaural line at a point that defines lambda. The mediolateral coordinates are defined relative to the midpoint of segment "c." This construct is a rule of thumb that applies equally well to both the rat and mouse. It is critical to the extent that it affects the determination of the dorsoventral coordinate and thus the horizontality of the skull.

5.10.1.1 Correction of Lambda

In some instances it may happen that lambda, as formally defined by the point at the base of the triangle formed by the intersection of the occipital and sagittal suture lines, is not located on the midline (Fig. 5.29). In this case, the ML coordinate of bregma can be used to correct for lambda such that the latter lies at the intersection of the midline (line "b") and the occipital suture line.

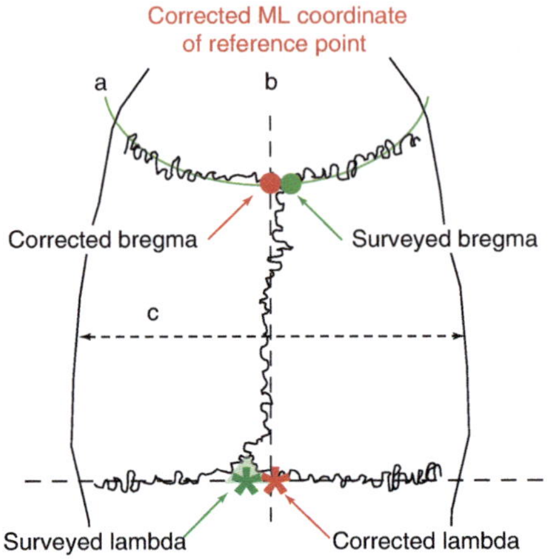

Fig. 5.30 Diagram showing the different bone suture lines in an animal with an asymmetric junction of the bone plates. In this example, lambda and bregma points are not aligned with the mediator (line "b"). Curve "a" represents the frontal sutures. The centerline "b" is equidistant from the lateral bone edges. This distance is designated by the segment "c"

5.10.1.2 Correction of Lambda and Bregma

In the example shown in Fig. 5.30, the two points are aligned on the horizontal plane, i.e., their dorsoventral coordinates are identical, but they are not aligned on the midline "b." In this case, the ML coordinate of each of the two points is corrected to be found on the midline.

Note: Occasionally, some animals are encountered with sutures or skull shapes that present too many anomalies for identifying landmarks with any certainty and therefore not exploitable for locating the target structure. It is preferable in this case to discontinue the procedure, to close the wound and attempt to reassign the animal for use in another protocol.

5.10.1.3 Correction of Bregma

In the example shown in Fig. 5.31, the surveyed lambda point situated on the centerline "b" is not in line with the bregma point surveyed on the frontal suture. In this case, the ML coordinated of bregma is corrected to correspond to that of lambda.

5.10.2 When the Dorsoventral Coordinates of Lambda and Bregma Are Different

If the operator finds a difference between the dorsoventral coordinate of the lambda and bregma landmarks, the orientation of the head of the animal should be adjusted so that this difference is less than or equal to 0.1 mm. If this verification is not respected, the attainment of the target structure, whose coordinates depend on a perfectly horizontal stereotaxic plane, is compromised. If this verification is not done and an eventual problem not corrected, it may recur in subsequent procedures

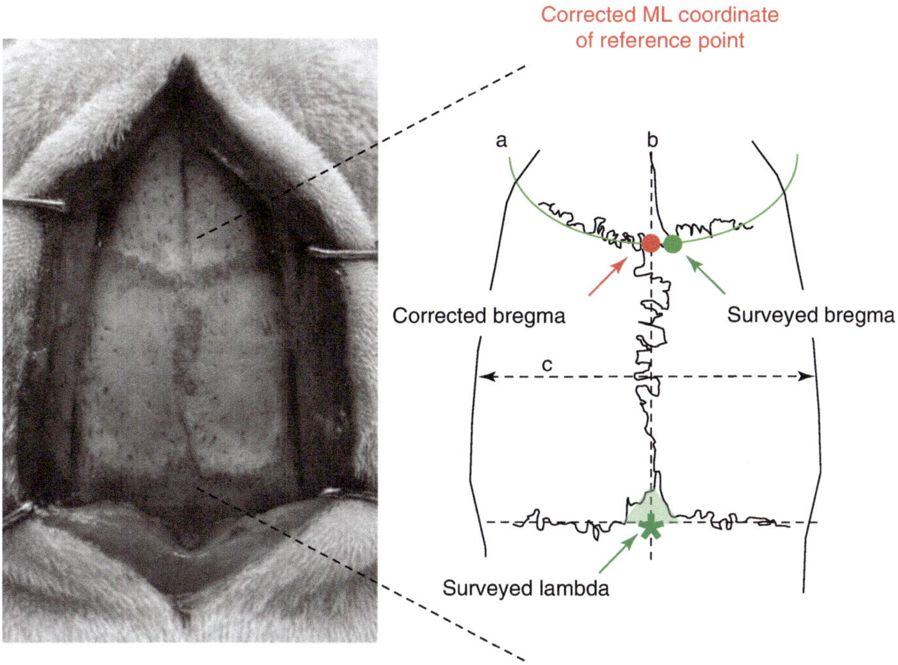

Fig. 5.31 Diagram showing the skull bone suture lines in an animal with an asymmetric junction of the lambdoid and frontal bone plates. In this example, the ML coordinate of lambda, situated on the centerline "b," will be used for bregma. The curve "a" represents the frontal sutures. The centerline "b" is equidistant from the two lateral bone edges. This distance is defined by segment "c"

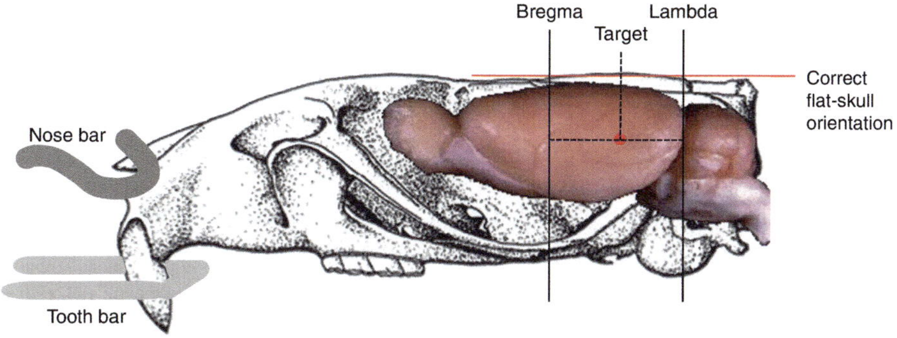

Fig. 5.32 Diagram showing the position of the virtual intracerebral target (*red point*) in an animal whose skull is correctly placed in the "flat-skull" position

and thus risk the exploitation of the experimental data obtained from these operated animals. A rigorous postmortem histological examination will reveal any discrepancy between the stereotaxic target and the endpoint of the descent actually achieved.

- If the dorsoventral coordinates of lambda and bregma have a difference of less than 0.1 mm, it is considered that the head is in the proper horizontal or "flat-skull" position (Fig. 5.32).

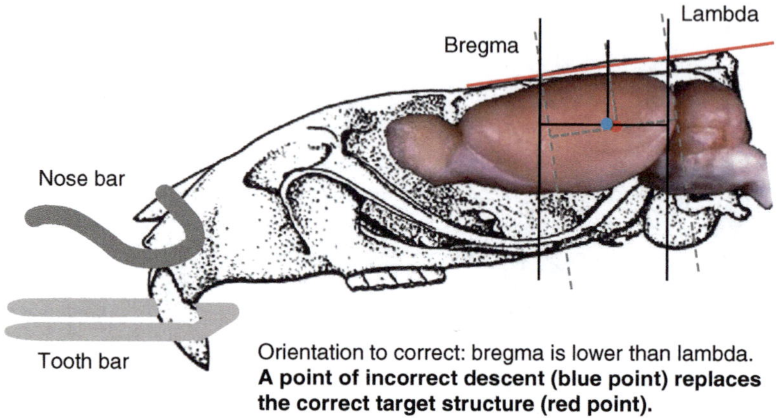

Fig. 5.33 Diagram showing the angle of inclination of the head of the animal and the incorrect placement (*blue point*) of the virtual intracerebral target (*red point on gray dashed line*) when the skull is not placed in the correct horizontal "flat-skull" position

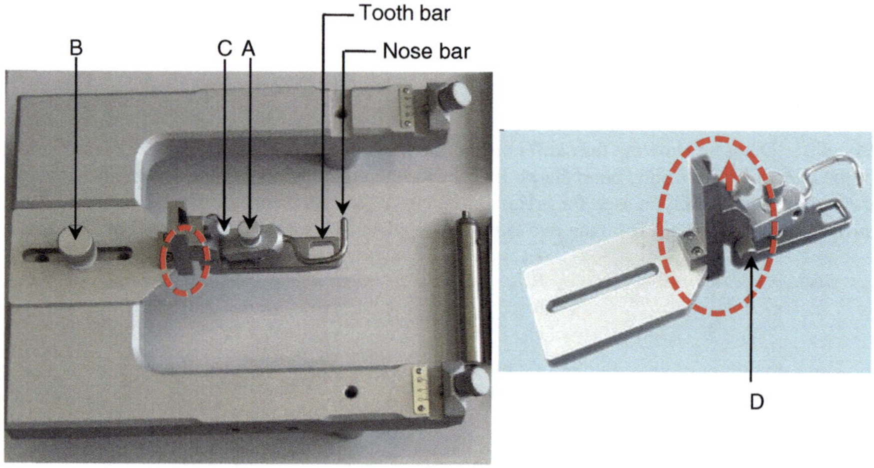

Fig. 5.34 The different setscrews of the nose and tooth bars

- When the dorsoventral coordinate of bregma is lower than lambda (and vice versa), the target does not lie along the correct vertical axis of descent (Fig. 5.33).

In order to realign the head of the animal correctly, the nose bar setscrew (A in Fig. 5.34), the setscrew of the tooth bar plate (B in Fig. 5.34), and the setscrew holding the vernier of the tooth bar (C in Fig. 5.34) are first loosened.

Since the dorsoventral coordinate of bregma is greater than and therefore situated lower than lambda, the tooth bar (D in Fig. 5.34) has to be raised to upwardly reorient the rostral part of the animal's head. After completing this manipulation,

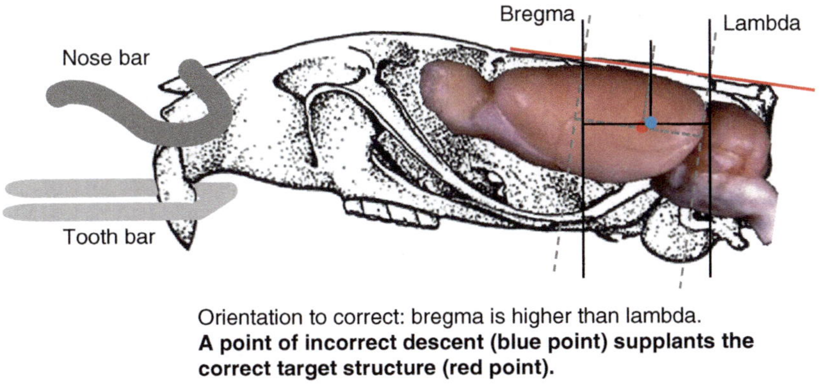

Fig. 5.35 Diagram showing the angle of inclination of the head of the animal and the incorrect positioning (*blue point*) of the virtual intracerebral target (*red point*) when the skull is not placed in the correct horizontal "flat-skull" position (*gray dashed line*)

the three setscrews are retightened in reverse order: the tooth bar vernier, the tooth bar plate, and the nose bar. The dorsoventral coordinates of both lambda and bregma are then surveyed again to verify that they are similar. This operation is repeated until the difference between the coordinates is less than or equal to 0.1 mm.

- When the dorsoventral coordinate of bregma is smaller than that of lambda, the target does not lie in the correct vertical axis of descent (Fig. 5.35).

After loosening the nose and tooth bar as in the previous example, it is necessary to lower the tooth bar (Fig. 5.36) in order to redirect the rostral part of the head of the animal downward.

At the end of the manipulation, the three setscrews are retightened in reverse order: the tooth bar vernier, the tooth bar plate, and the nose bar. The dorsoventral coordinates of both lambda and bregma are then surveyed again to verify that they are similar. This operation is repeated until the difference between the coordinates is less than or equal to 0.1 mm. The use of a binocular loupe is sometimes necessary to carry out this operation with precision.

Note: The rigor and quality of this procedure have an important impact on the results obtained, since the slightest inaccuracy can lead to poor stereotaxic positioning. This could invalidate the results from one or more animals, which is ethically unacceptable.

When all of the procedures described above are respected, the calculation of the target structure coordinates can be carried out and recorded in the animal's monitoring worksheet (Appendix 2A). The operator can then proceed with the implantation itself.

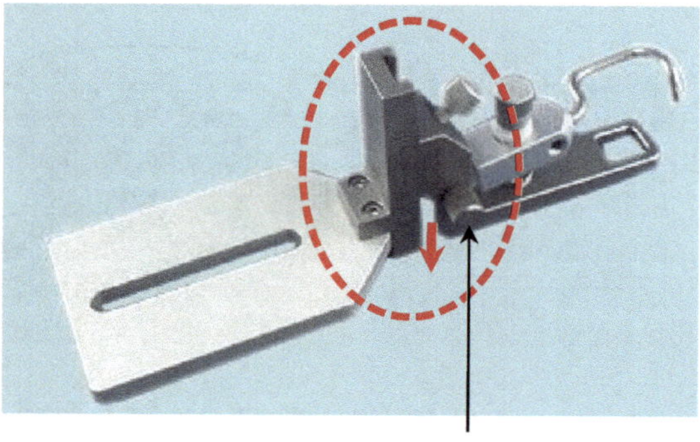

Fig. 5.36 Anterior part of the stereotaxic apparatus showing the descent (*arrow*) of the dorsoventral vernier of the tooth bar

These steps constitute the basic surgical approach to targeting most intracerebral structures. The operator can then proceed with the specific implantation phase of his experimental protocol.

5.11 Suturing

Depending on the type of surgery performed, the wound may or may not need to be sutured at the end of the procedure. The means and methods of suturing have advanced significantly over the past decades, and sterile ampoules of catgut and surgical cotton have lead to a revolution of postoperative developments in surgical interventions. Today the surgeon has an array of different suturing materials (Appendix 9) and biocompatible adhesives at his disposal to tackle a diversity of indications. Surgical suturing threads, by their multiple physical and chemical characteristics together with the various types of needles associated with them, can respond to most needs in experimental neurosurgery. Suturing is generally carried out using a special thread crimped onto a sterile needle. The purpose of a suture is to bring together and maintain two edges of a wound in order to obtain rapid healing. It minimizes inflammation and complications related to infection and spontaneous wound opening. When performing the surgical incision, the operator should take care to preserve a good vascularization of the incised tissue and to master the suturing techniques to assure a good approximation of all tissue layers, deep and superficial, without excessive force and with the proper type of suture thread. As described previously for the surgical procedures, aseptic conditions ensure an environment favorable to wound healing.

The ideal suture thread has specific biological and physical characteristics: it must be sterile, biocompatible and tissue resistant, of minimal capillarity and high tensile strength, and with excellent knot-holding properties and easy handling and a perfectly predictable and reproducible behavior.

The packaging of surgical sutures (Fig. 5.4) in individual blister packs shows all the relevant product characteristics. Once chosen, the operator should check the conformity of the sutures at his disposition before starting the surgical procedure.

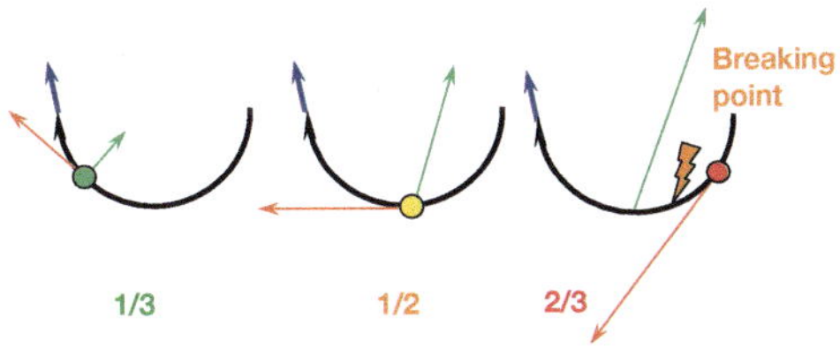

Fig. 5.37 Diagrams showing the angles of force made by the needle as a function of the position of the needle holder

Different types of suture threads and needle mountings (Table 5.1) are often used during any one procedure. While it is not possible to keep an opened suture needle for more than one operation in a human patient, they can be reused in a series of experimental procedures in animals, if the animals are healthy and microbe free. As for the surgical instruments, suture materials unsoiled by body fluids are protected from contamination by a sterile drape, to avoid contamination when changing from one animal to the next.

In order to avoid modifying their coating and thus their capillarity, as well as their strength and/or resorption characteristics, sutures must neither be washed nor disinfected before reuse. For effortless suturing, a disproportionate solicitation of the needle should be avoided by positioning the needle holder at 1/3 of the needle length (from the point) as shown in Fig. 5.37.

5.11.1 *Surgical Staples*

Although staples are easier and faster to insert and the probability of secondary fistulas are identical to those of a conventional suture, their use entails higher risks of early bleeding and secondary stenosis. Made of steel or titanium alloy, they are not absorbable and once placed should be forecasted for eventual removal. This operation is sometimes complicated since they can become firmly lodged within the tissue, producing areas where vascularization is minimal or absent.

5.11.2 *Biocompatible Glues and Adhesive Strips*

Adhesive strips, commonly used on humans to close small superficial wounds or to complete an intradermal suture, can be useful for protecting a wound and consolidating sutures in small animals such as rodents. They should be applied without excessive tension that might cause compression and local blood hypoperfusion. Cyanoacrylate-based biological adhesives (e.g., Dermabond®) allow rapid closure of skin. Because of the efficient sealing properties of such products, they are reserved for traumatic or surgical wounds that are perfectly clean, clear, and not under tension. Their use in routine experimental procedures is somewhat limited by their relatively high cost.

5.11.3 Making a Simple Stitch

A review of the general principles governing the tying of all knots and applying to all suture materials can be found in the *Knot Tying Manual* (Ethicon 1986), *Covidien Surgical Knot Tying Manual* (2008), and similar publications freely available online.

Generally, a braided (multifilament) thread is preferred for interrupted sutures, since its flexibility and elasticity favor an even tension and durable surgical knots. The following steps are followed in performing a simple suture (Fig. 5.38).

With a needle held at its distal third using a needle holder, one edge of the wound is pierced (Fig. 5.38-1). The distal portion of the needle is pushed through, respecting its curvature (Fig. 5.38-2). The needle is then released and grasped once again with the needle holder between the two edges of the wound. This operation is repeated for the opposite edge of the wound (Fig. 5.38-3). By avoiding to take on both edges of the wound at once, the surgeon can adjust the tension of the thread on each side to compensate for any slight asymmetry between the two wound edges (Fig. 5.38-4, -5).

The needle holder, positioned between the two ends of the suture thread, is wrapped once by the lead (needle) strand, shown in turquoise in Fig. 5.38-6, and then picks up the short tail strand shown in dark blue in Fig. 5.38-7. Dragging the

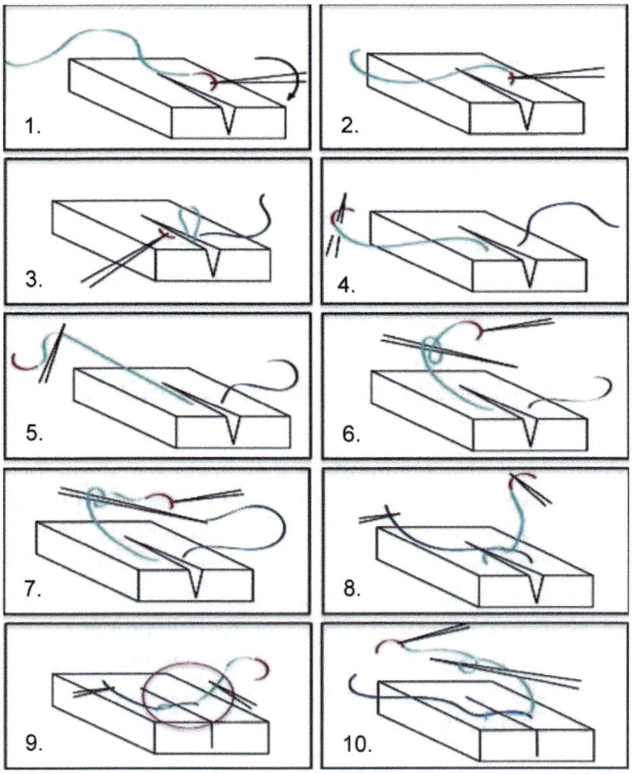

Fig. 5.38 The different steps in performing a simple suture

loop of the long lead strand towards the wound, a simple single overhand throw is established (Fig. 5.38-8). The operation is repeated to make a second overhand throw in the reverse sense (Fig. 5.38-9, -10), again with the needle holder positioned between the two ends of suture thread. The opposition of these two overhand throws forms a "square" knot that prevents thread slippage.

To make a so-called surgeon's knot, three overhand throws are performed: the first for tightening and confirming the approximation of the divided wound edges, the second for adjusting and locking the previous throw, and the third for insurance (Fig. 5.39).

It is also possible to do a double throw in place of the first simple one, in order to impose a slight tension between the two wound edges and to reduce any slippage of the suture thread. The needle holder, positioned between the lead and tail strands of thread, is wrapped *twice* by the lead (needle) end of the thread instead of only once.

For monofilament sutures, two or three additional throws must be made to compensate for the higher slippage characteristics of the thread. Interrupted sutures are individually solid: if one loosens, there is no serious consequence for the others. However, the tightness of each one must be carefully controlled in order to distribute the tension of the closure over the entire extent of the wound.

5.11.4 Running Sutures

For a running suture (Fig. 5.40), a monofilament with substantial slippage optimizes the distribution of passive tension. A knot of multiple throws at each end of the closure does not affect the rapidity of its execution. Running sutures, involving several successive passages over the entire length of the wound closure, partially offsets any irregularity in the thread tension, but if one of the end knots loosens, the solidity of the whole suture is compromised. A simple running suture consists of a single stitch where only the tail strand of thread is cut short, followed by successive passes of the lead strand always in the same direction. The last pass in the opposite direction forms a loop that takes the place of the tail strand in making the terminal surgeon's knot. A variation of this last step is to use a metal clip or a "button" to hold the two ends of the running suture, thereby facilitating its removal.

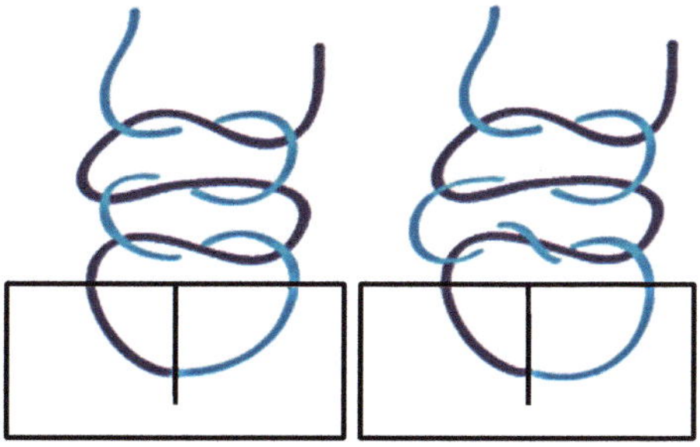

Fig. 5.39 Overhand throws performed with monofilament thread (Appendix 9)

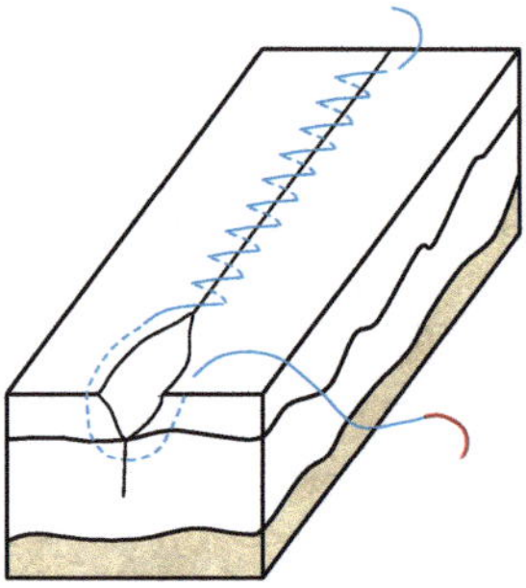

Fig. 5.40 Running suture combining several successive passes of the suture thread along the entire length of the wound

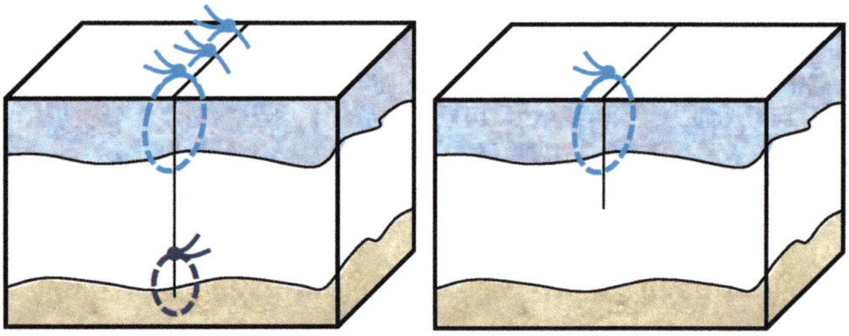

Fig. 5.41 Suturing of one or more tissue layers, depending on the depth of the incision

5.11.5 The Keys to a Good Suture

An important part of good suturing technique is correct method in knot tying. A seesaw motion, or the sawing of one strand down over another until the knot is formed, may materially weaken sutures to the point that they may break when the second throw is made or, even worse, in the postoperative period when the suture is further weakened by increased tension or motion. If the two ends of the suture are pulled in opposite directions with uniform rate and tension, the knot may be tied more securely.

The closure of a surgical wound is intended to restore the anatomical integrity of transected tissues, layer by layer (Fig. 5.41). While ensuring a good approximation of the tissue edges from the deep to superficial parts of the wound, the operator must make sure not to leave behind any foreign body, pieces of thread or cotton gauze, or empty cavities that may lead to fluid collection. If such a pocket forms, it could gradually fill with a sero-hemorrhagic fluid. This warm closed space represents an ideal breeding ground for bacteria. An imperfect suture can thus produce a pocket of pus that must eventually be treated, otherwise risking the loss of the animal.

To obtain the smallest possible and least debilitating scar, the suturing of the superficial layer, skin, or mucous membrane should be particularly neat for perfect approximation of the wound edges (Fig. 5.42) and less risk of inflammation or infection.

After suturing, the wound should be carefully cleaned with a compress soaked with a 1:10 dilution of disinfectant. Any traces of glue, cement, hair, or blood clots that can

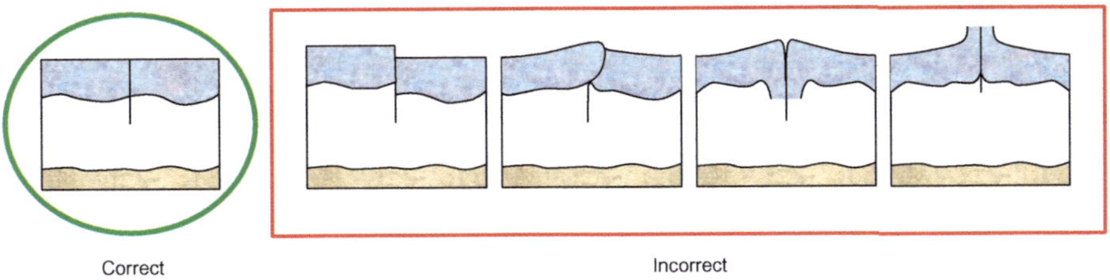

Fig. 5.42 Examples of tissue layer approximation after suturing

interfere with wound healing are carefully removed. Before the animal awakes from anesthesia, the nails on all four paws are carefully trimmed to minimize any impact of the animal's normal scratching and grooming behavior on the healing process.

> *Note*: For certain implantation procedures, suturing may not necessarily be required at the end of the operation, e.g., in acute experiments of short duration where postoperative recovery of the animal is not intended.

5.12 Postoperative Follow-up of the Animal

When surgery is completed, the animal is placed on absorbent paper in a cage to limit contact of the suture with dust or litter chips. Hypothermia should be treated for as long as the anesthesia lasts. An appropriate temperature in the recovery room should be maintained, especially in the cage, either by means of an incubator, a heating pad, or a lamp. Some moistened food pellets are placed within the cage to facilitate feeding.

The animals are returned to the housing facilities only after complete awakening from anesthesia.

From a general point of view, a good monitoring procedure can prevent complications occurring after surgery. Frequent observation is required to identify any discomfort or pain caused by surgical trauma and to remedy it when necessary. Cage bedding may have to be changed more frequently if it has been soiled by bloody discharges. If necessary, during daily examinations, sutures can be cleaned using a sterile cotton swab moistened with disinfectant. All metal- and/or antibiotic-based sprays should be avoided since they can introduce foreign microbodies that can lead to increased inflammation at the wound site. In the case of significant bleeding, direct pressure can be applied for a few minutes. If it persists, the animal should be reanesthetized for proper management of hemostasis. In the case of suture opening, the wound can become contaminated and must be treated as a traumatic lesion. Antibiotic treatment can then be established by the veterinarian to preserve the normal evolution of the healing process. In the case of pain, reinjection of the appropriate analgesic drugs can be used while deciding on the fate of the animal. This will depend on the severity of the complication and the compatibility of the treatment with the nature of the experiment.

If the operator chooses to administer an analgesic by the oral route by including it in the animal's drinking water, the fluid consumption must be monitored. Since anesthesia can cause physiological disturbances of different severity, the animal may have difficulty accessing the source of drinking water and/or simply be less inclined to eat or drink during the immediate postsurgical period. Since this effect is independent of the invasiveness of the surgery, it is best to administer the analgesic parenterally for at least 12–24 h, until the animal's state allows it to be autonomous in its feeding and drinking behavior. An energy supplement (e.g., Ovomaltine®) in the drinking water or any other palatable food source may also be necessary to avoid excessive weight loss. *In the case of anorexia or any suspected pathology, lack of grooming, or abnormal posture, the corrective action to take will be determined solely by the attendant veterinarian.* If all the choices made for the implementation of intracranial surgery are consistent and well conducted, the animal should have no unanticipated postoperative complications. Despite all the precautions taken, in the case of an original abnormal phenomenon, the principal investigators should be questioned about the experimental model used and be referred to the veterinarian.

5.13 Care and Storage of Instruments

After each operation, the entire stereotaxic apparatus and the reusable instruments used in the procedure should be thoroughly cleaned, decontaminated, and eventually sterilized. Decontamination begins with a basic soap and water wash. When all the materials are clean, including specific hardware and trepanation equipment, they should be dried thoroughly before autoclaving. Materials used for implantation, cannulae, obturators, electrodes, etc., are either decontaminated or autoclaved depending on their particular technical tolerances. The entire stereotaxic device should be thoroughly cleaned and decontaminated. For this, cotton swabs or paper towels soaked with an antiseptic solution of 70 % ethyl alcohol are applied to the frame, insisting on the nose and tooth bars and also the verniers. This same cleaning procedure can be used for the heating pad, rectal temperature probe, and workbench.

Note: before storing the equipment, the operator must ensure that the vernier graduation markings located on the fixed arm of each micromanipulator lie within the scale range from 0 to 8 (0 to 80 units) on the movable arm (Fig. 5.43).

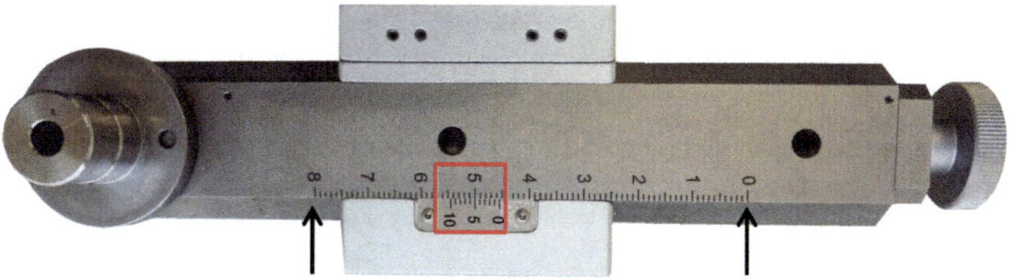

Fig. 5.43 Vernier located on the anterior-posterior axis of the stereotaxic apparatus. The *gray box* showing the vernier digits must always be situated between the *two arrows* that indicate the decimal scale markings from 0 to 80 units

Finally, depending on the frequency of use of the stereotaxic devices, preventive maintenance, checkups, and servicing of the equipment should be carried out regularly. This will ensure a precise fit and adjustment of the micrometric screws and prevent stereotaxic error resulting from any eventual looseness or "play" in these elements.

References

Bekker A, Sturaitis MK (2005) Dexmedetomidine for neurological surgery. Neurosurgery 57(1 Suppl):1–10

Blasiak T, Czubak W, Ignaciak A, Lewandowski MH (2010) A new approach to detection of the bregma point on the rat skull. J Neurosci Methods 185(2):199–203

Chery R, Gurden H, Martin C (2012) Anesthetic regimes differentially modulate the spontaneous and odor-induced temporal dynamics in the mouse olfactory bulb. (Submitted)

Drummond J (2000) Monitoring depth of anesthesia, with emphasis on the application of the bispectral index and the middle latency auditory evoked response to the prevention of recall. Anesthesiology 93:876–882

Edlich RF, Long WB (2008) Covidien surgical knot tying manual. Covidien, Norwalk, p 95, www.covidien.com

Ethicon, Inc (1986) Knot tying manual. Ethicon, Somerville, p 40, http://www.sutureonline.com/suture-knot-tying-manual

Farber NE, Harkin CP, Niedfeldt J, Hudetz AG, Kampine JP, Schmeling WT (1997) Region-specific and agent-specific dilation of intracerebral microvessels by volatile anesthetics in rat brain slices. Anesthesiology 87(5):1191–1198

Ferrante FM, Paggioli J, Cherukuri S, Arthur GR (1996) The analgesic response to intravenous lidocaine in the treatment of neuropathic pain. Anesth Analg 82:91–97

Fornari RV, Wichmann R, Atsak P et al (2012) Rodent stereotaxic surgery and animal welfare outcome improvements for behavioral neuroscience. J Vis Exp (59). doi:10.3791/3528

Harrison NL (2003) Knockin' on the door of general anesthetic mechanisms: but will US researchers be shut out ? Anesth Analg 97:616–618

Hemming HC Jr (2009) Sodium channels and the synaptic mechanisms of inhaled anaesthetics. Br J Anaesth 103(1):61–69

Jensen EW, Litvan H, Struys M, Vazquez M (2004) Pitfalls and challenges when assessing the depth of hypnosis during general anesthesia by clinical signs and electronic indices. Acta Anaesthesiol Scand 48:1267–1269

Joris J (2008) La lidocaïne par voie intraveineuse. Revue Médicale de Lièges 63:2–9
Jurd R, Arras M, Lambert S et al (2003) General anesthetic actions in vivo strongly attenuated by a point mutation in the GABA(A) receptor beta3 subunit. Off Publ Fed Am Soc Exp Biol 17:250–252
Koppert W, Ostermeier N, Sittl R, Weidner C, Schmelz M (2000) Low-dose lidocaine reduces secondary hyperalgesia by a central mode of action. Pain 85:217–224
Kuno M, Matsuura S (1982) Sites and mechanisms of action of lidocaine upon the isolated spinal cord of the frog. Brain Res 249:87–93
Li X, Wu J, Zheng J, Li Y, Yang T, Hu G, Dai J, Yang Q, Dai L, Jiang Y (2013) Altered miRNA expression profiles and miR-1a associated with urethane-induced pulmonary carcinogenesis. Toxicol Sci 135(1):63–71
Mason RJ, Kalina M, Nielsen LD, Malkinson AM, Shannon JM (2000) Surfactant protein C expression in urethane-induced murine pulmonary tumors. Am J Pathol 156:175–182
Messier C, Emond S, Ethier K (1999) New techniques in stereotaxic surgery and anesthesia in the mouse. Pharmacol Biochem Behav 63(2):313–318
Nagy I, Woolf CJ (1996) Lignocaine selectively reduces C fibre-evoked neuronal activity in rat spinal cord in vitro by decreasing *N*-methyl-D-aspartate and neurokinin receptor mediated post-synaptic depolarizations; implications for the development of novel centrally acting analgesics. Pain 64:59–70
Nelson LE, Guo TZ, Lu J, Saper CB, Franks NP, Maze M (2002) The sedative component of anesthesia is mediated by GABA(A) receptors in an endogenous sleep pathway. Nat Neurosci 5:979–984
Paxinos G, Watson C (2004) The rat brain in stereotaxic coordinates, 5th edn. Academic, San Diego
Richardson CA, Flecknell PA (2005) Anaesthesia and post-operative analgesia following experimental surgery in laboratory rodents: are we making progress ? Altern Lab Anim 33(2):119–127
Shavit Y, Fish G, Wolf G et al (2005) The effects of perioperative pain management techniques on food consumption and body weight after laparotomy in rats. Anesth Analg 101:1112–1116
Smith DJ, Perrotti JM, Mansell AL, Monroe PJ (1985) Ketamine analgesia is not related to an opiate action in the periaqueductal gray region of the rat brain. Pain 21(3):253–265
Sonner JM, Antognini JF, Dutton RC et al (2003) Inhaled anesthetics and immobility: mechanisms, mysteries and MAC. Anesth Analg 97:718–740
Stokes EL, Flecknell PA, Richardson CA (2009) Reported analgesic and anaesthetic administration to rodents undergoing experimental surgical procedures. Lab Anim 43(2):149–154
Tanelian DL, MacIver MB (1991) Analgesic concentrations of lidocaine suppress tonic A-delta and C fiber discharges produced by acute injury. Anesthesiology 74:934–936
Vizcaychipi MP, Lloyd DG, Wan Y, Palazzo MG, Maze M, Daqing M (2011) Xenon pretreatment may prevent early memory decline after isoflurane anesthesia. PLoS One 6(11):1–7
Waite A, Gilliver SC, Masterson GR, Hardman MJ, Ashcroft GS (2010) Clinically relevant doses of lidocaine and bupivacaine do not impair cutaneous wound healing in mice. Br J Anaesth 104(6):768–773
Wallace MS, Dyck JB, Rossi SS, Yaksh TL (1996) Computer-controlled lidocaine infusion for the evaluation of neuropathic pain after peripheral nerve injury. Pain 66:69–77

Chapter 6
Realization of the Stereotaxic Surgery

6.1 Generalities

Stereotaxic surgery is used in combination with other diverse techniques necessary for addressing the specific objectives of a proposed study. These techniques include experimental procedures in vivo for activating or inactivating brain structures or transmitter systems, permanent selective lesioning, functional neuroanatomy using tracers, and acute or chronic measurements or recordings. The principal aspects of some of these techniques are described below.

Stereotaxic neurosurgery can be used in lesioning (damaging, depleting, or ablating) a structure or transmitter system:

- By electrolysis (Duncan et al. 1975)
- By transection (Milner and Amaral 1984)
- By aspiration (Berridge and Whishaw 1992; Goldstein 1993; Vargo and Marshall 1996)
- By radio frequency (King 1991)
- By injection of excitotoxins such as ibotenic acid (Jarrard 2002), 6-hydroxydopamine for dopaminergic neurons (Rotman and Creveling 1976; Kostrzewa 1995, Kostrzewa and Brus 1998; Schwarting and Huston 1996, 1997; Glinka et al. 1997; Centonze et al. 1999; Blum et al. 2001), *N*-(2-chloroethyl)-*N*-ethyl-2-bromobenzylamine (DSP4) for noradrenergic neurons (Jonsson 1980; Fritschy and Grzanna 1991; Dudley et al. 1990; Kostrzewa 2009) and immunotoxins such as 192 IgG-saporin for cholinergic systems (Wiley 1992, 1997, 2001; Schliebs et al. 1996; McGaughy et al. 2000)
- By global or focal ischemia using devascularization, photo-irradiation, vessel occlusion, or vasoconstrictive agents (see Livingston-Thomas et al. 2013 and references therein)

Stereotaxic neurosurgery can permit the temporary inactivation of a structure:

- By microinjection of muscimol (GABA agonist), tetrodotoxin (TTX), or lidocaine (to block nerve conduction by reducing membrane permeability of sodium ions that occurs during depolarization) (Chen et al. 2000)
- By microinjection of receptor antagonists of the target system (see Chap. 2)

B. Ferry et al., *Stereotaxic Neurosurgery in Laboratory Rodent*,

DOI 10.1007/978-2-8178-0472-9_6,

Stereotaxic neurosurgery can be used in stimulating a CNS structure:

- Electrically (nonselective activation of a heterogeneous population of neurons)
- By microinjection of *N*-methyl-D-aspartate (NMDA) that stimulates neurons bearing NMDA receptors
- By microinjection of receptor agonists of the target system (see Chap. 2)
- By applying a light stimulus (laser) after transfection of modified gene sequences (e.g., channel rhodopsin-2; Buchen 2010)

Stereotaxic neurosurgery can be used when marking a CNS structure:

- By retrograde labeling, e.g., with horseradish peroxidase (HRP) or fast blue (Milner and Amaral 1984)
- By anterograde labeling, e.g., with *Phaseolus vulgaris* leukoagglutinin (PHA-L) or radiolabeled amino acids (Milner and Amaral 1984)

Stereotaxic neurosurgery is also used in making in vivo *measurements of:*

- The concentration of extracellular neuromediators (by microdialysis and voltammetry; Li et al. 2006)
- Electric potentials (by electrophysiology)

6.2 Insertion of a Guide Cannula

This technique is used when manipulating the activity of a structure in a conscious or anesthetized rodent in acute or chronic experimental situations. A unilateral implantation into the dorsal hippocampus is used here as an example.

> The stereotaxic coordinates of the target are:
>
> AP, – 3.14; ML, 2.0; and DV, – 3.5 from the dura mater in the rat
>
> AP, – 1.8; ML, 1.5; and DV, – 2.4 from the dura mater in the mouse
>
> Guide cannula: stainless steel; 7 mm long; 0.3 mm internal diameter and 0.4 mm external diameter for both the rat and the mouse
>
> Anchoring screws: stainless steel; length 1.20 mm and diameter 1.85 mm

The unilateral implantation of the guide cannula is performed under general anesthesia. After having exposed the cranial surface and before trepanation, tissue retractors (see Fig. 5.3) are placed on either side of the cranial area of interest (Fig. 6.2a) and

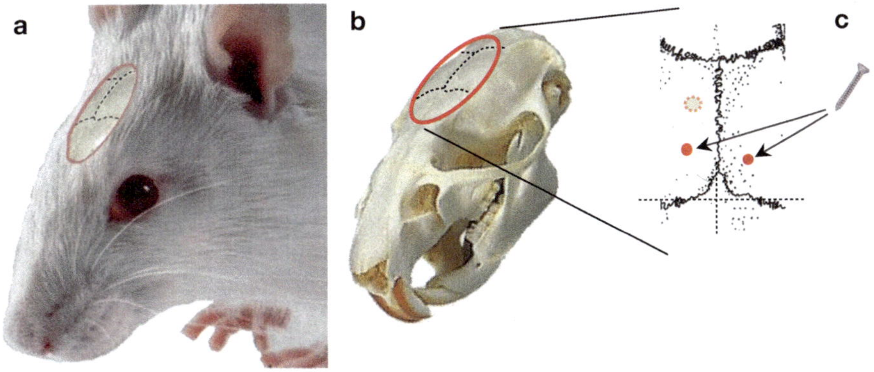

Fig. 6.1 Anchor screw placement. (**a**) Diagram showing the sutures of the cranial bones in the animal's head. (**b**) Corresponding view of the animal's skull. (**c**) Enlargement of the area corresponding to bregma and lambda reference points. The small *dashed circle* represents the location where the cannula will be inserted according to the coordinates for the dorsal hippocampus (in this example). The *two red points* correspond to the approximate locations of two anchoring screws

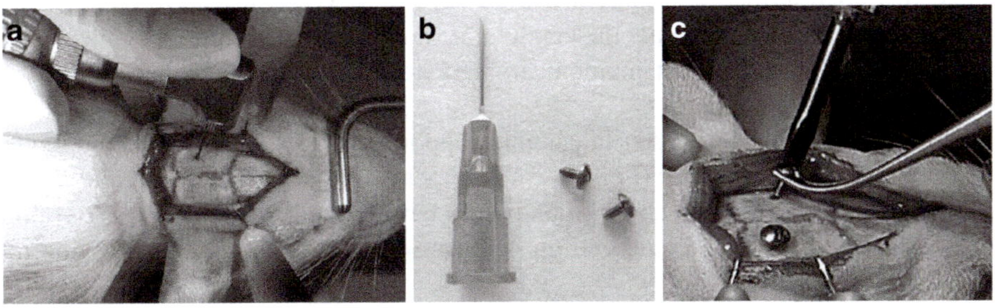

Fig. 6.2 Mounting of the anchor screws following the preparation of the operatory zone. (**a**) Drilling of pilot holes for placement of the screws. (**b**) Image of anchor screws and a Terumo 25G hypodermic syringe needle showing their comparative sizes. (**c**) Insertion of anchor screws

anchor screws are attached to the skull. For this, pilot holes with a diameter slightly smaller than that of the screw thread (ideally the screw's core diameter) are drilled about 3 mm distant from the desired point of insertion of the cannula.

For most procedures in the rat, two or three screws are used. For the mouse, one or two are usually sufficient (Fig. 6.1). For proper insertion, care should be taken to leave a space of at least 1 mm between the screw head and the surface of the skull. Care should also be taken that the screws are not inserted so far as to touch the surface of the brain.

Once the anchor screws are inserted (Fig. 6.2), the stereotaxic coordinates are taken. The horizontality of the skull is carefully verified, the emplacement of the guide cannula pinpointed, and the insertion hole drilled (Fig. 6.3a).

6.2.1 Use of the Dental Drill and Handpiece

When the DV coordinate at the point situated on the vertical axis of the target has been surveyed and identified as described in the previous paragraph, the skull bone is drilled to access the dura. For this, a dental drill with drill bit and handpiece that have been previously sterilized are usually used. After the motor is turned on, the drill bit is applied to the skull surface applying just enough force to pierce the bone. This gesture should be carried out as quickly as possible to avoid generating heat caused by friction of the rotating bit, which can be further reduced by simultaneous irrigation with sterile saline. *Particular care should be taken regarding the amount of force applied while drilling.*

If too much force is applied and the bone is penetrated too deeply, there is a risk of damaging the dura and the surface of the cortex. After trepanation, the dura is delicately hooked using the tip of a small sterile hypodermic needle and then incised in a circular motion. The correct execution of this step is evident by the appearance of cerebrospinal fluid seeping out to the surface of the skull.

To facilitate the introduction of the cannula into the brain, the dura mater is perforated, or a small surface of it resected, using a sterile beveled syringe needle (Fig. 6.3b).

In order not to contaminate the cannula, it should be installed on the object carrier and inserted at the last moment, immediately after the dura mater has been perforated.

After having installed a sterile cannula support or holder, which fixes the guide cannula to the object carrier (Fig. 6.4), it is imperative to check the verticality of its axis relative to the coronal and median planes before it is lowered (Fig. 6.5). If this is not done, the cannula could end up being inserted at an angle, thus missing its intended target.

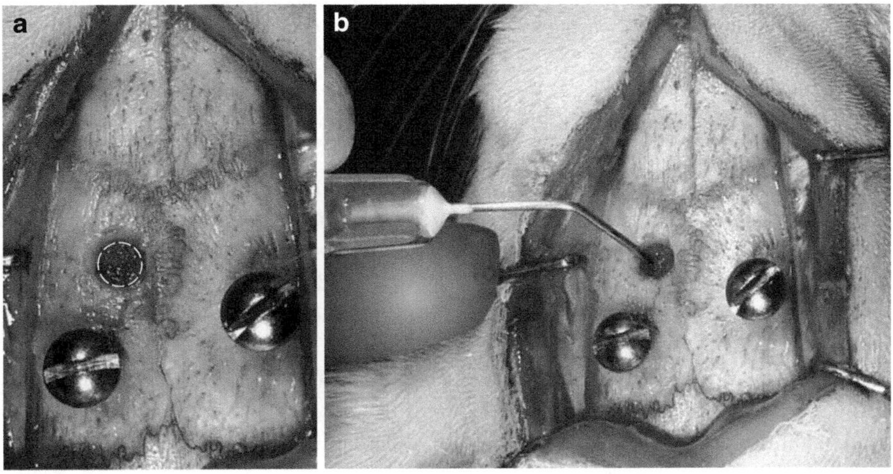

Fig. 6.3 (**a**) After the drilling of the insertion hole for the guide cannula. (**b**) Perforation of the dura mater using a sterile syringe needle to facilitate the insertion of the cannula guide

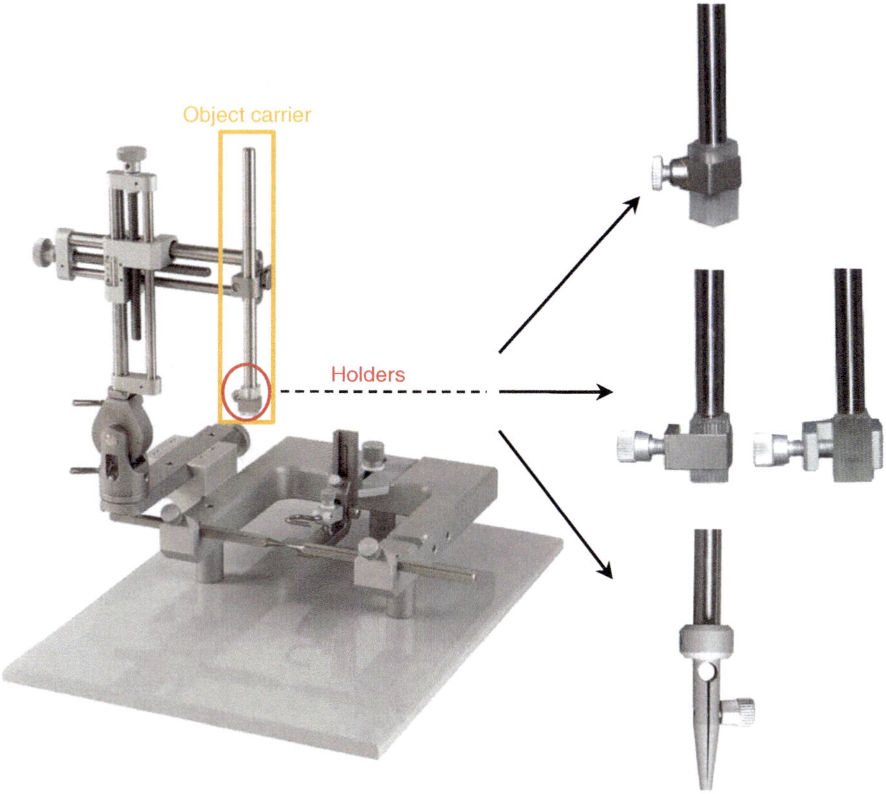

Fig. 6.4 Stereotaxic apparatus with various object (cannula, electrode, etc.) carriers that can be used for the type of implantation described above

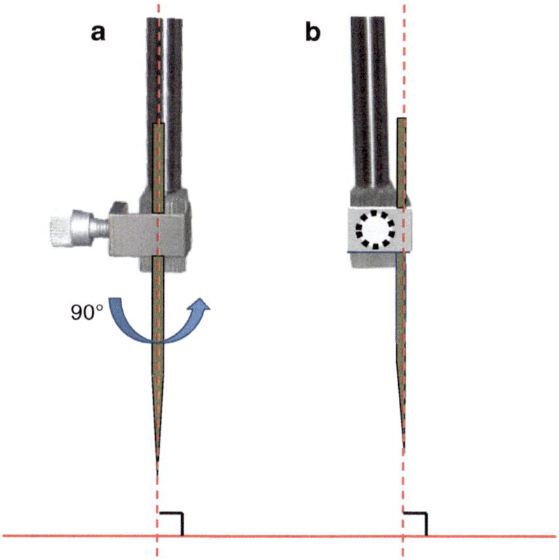

Fig. 6.5 Verification of the verticality of the instrument (e.g., cannula or electrode support) placed on the object carrier. (**a**) Coronal plane. (**b**) Median plane

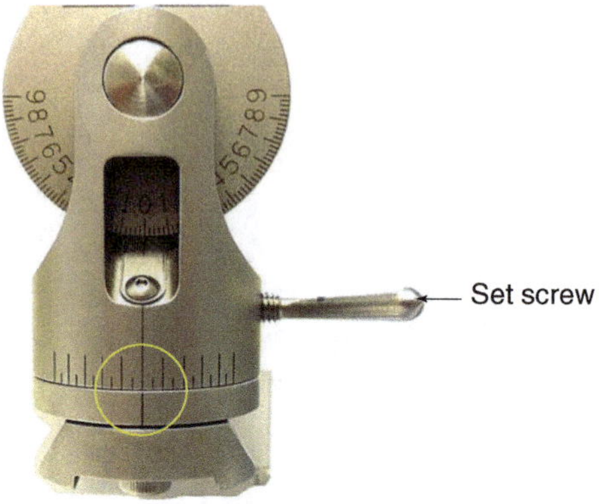

Fig. 6.6 Photograph showing the detail of the base of the vertical manipulator arm

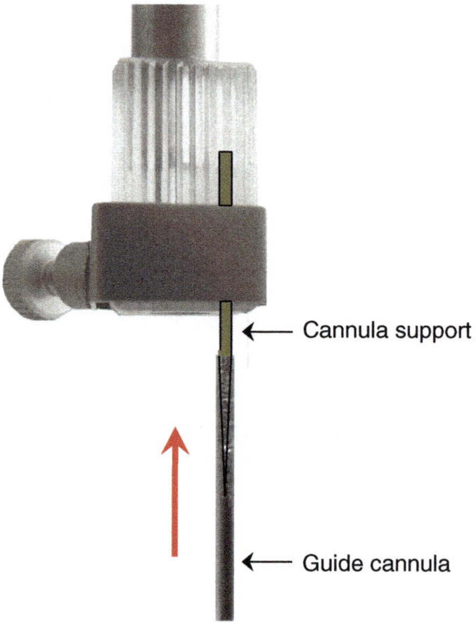

Fig. 6.7 Diagram showing the mounting of the guide cannula (in the direction of the *red arrow*) onto the cannula support

When verticality has been verified, the object carrier is fixed to the vertical stereotaxic manipulator arm. The tracking of the arm should be carefully checked with respect to the graduations marked on its base. For convenience, it is preferable to align the arm with the principle graduation mark as reference. Any rotation of the arm necessary for this alignment does not affect its verticality. The validity of this reference point is indispensable if it becomes necessary to set an angle for inserting the cannula (Fig. 6.6). The guide cannula that will remain permanently implanted in the brain is then mounted onto the cannula support for insertion (Fig. 6.7).

Next, the guide cannula is carefully lowered according to the predetermined dorsoventral coordinates (in this example, DV = −3.5 mm for the rat, or DV = −2.4 mm for the mouse), checking that its displacement is smooth and even along the dorsoventral axis (Fig. 6.8). This step is very important: if the orifice of the guide cannula is not centered correctly with respect to the target coordinates, the cannula risks

Fig. 6.8 Implantation of a guide cannula after placement of the anchor screws. Special attention should be paid to leave a space of about 1 mm distance between the head of the screw once fixed and the surface of the skull. The vertical red arrow indicates the direction followed by the object holder for the insertion of the guide

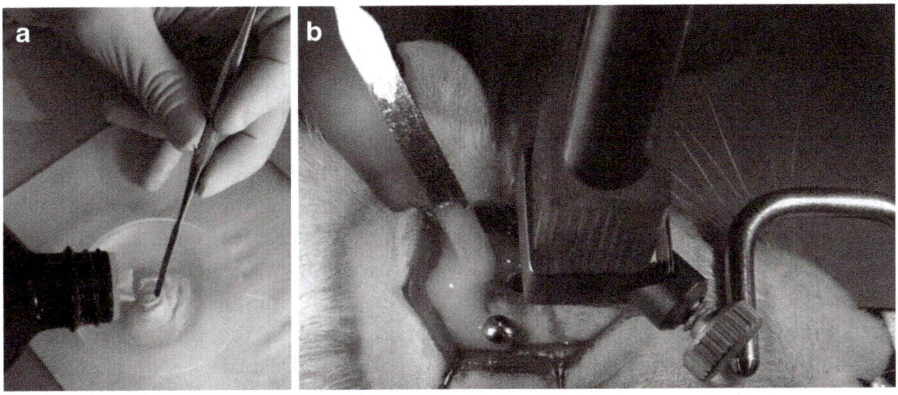

Fig. 6.9 Preparation of dental cement by bulk mixing (**a**) prior to its application onto the skull with a fine spatula. (**b**) Once the cement has entirely encircled the screws and cannula, it is necessary to wait several minutes to allow it to harden

touching the bone at the edge of the insertion hole, deviating its verticality and compromising an accurate trajectory towards the target.

The rest of the surgical operation may have different experimental goals requiring specific procedures. Some of these are described in the examples that follow in this chapter.

The guide cannula and anchoring screws must be fixed solidly to the skull of the animal using dental acrylic cement (e.g., Paladur® powder and solvent, Heraeus Kulzer & Co. GmbH, or Unifast Trad® powder and solvent, GC America Inc.) that can flow easily and finely to coat the guide cannula and anchor screws (Fig. 6.9).

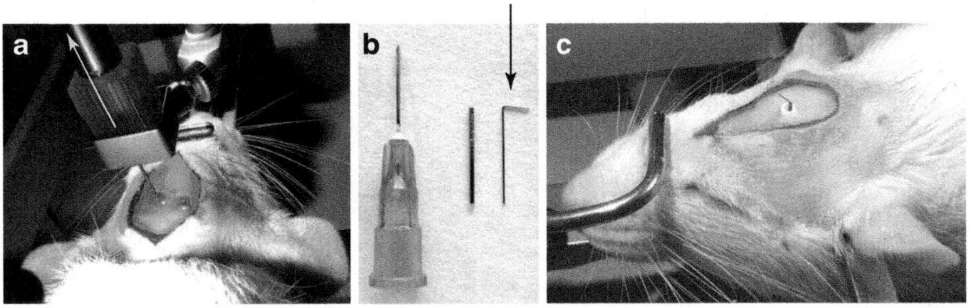

Fig. 6.10 (**a**) Ascent of the object carrier, in the direction of the *yellow arrow*. (**b**) Guide cannula and its obturator (*black arrow*) shown beside a 25G Terumo hypodermic showing their comparative sizes. (**c**) Appearance of the guide cannula implant at the end of the procedure

The cement should have a fluid consistency so that it acquires a maximum hardness after it sets. It is recommended to use aliquots of resin powder and solvent in vessels that should be kept sterile on the instrument tray so as not to break the chain of asepsis during preparation of the cement.

Before applying the dental cement, it is recommended to first apply a self-curing adhesive resin such as (Super-Bond C&B®, Sun Medical Co., Japan), especially if the skull is to be subjected to high mechanical stress (Gervasoni et al. 1998, 2000) or if the implant is required to hold for several months. This type of adhesive exists in the form of a monomer powder and a liquid polymer that are mixed together with a few drops of catalyst just prior to use. For optimal adhesion, it is necessary to etch the bone surface by using a cotton swab to apply a solution, such as mixture B in the Super-Bond® kit, containing 10 % citric acid. Because of its corrosive property, care should be taken not to prolong the contact of this solution with the bone for more than half a minute and to carefully dry the bone surface after its preparation. The Super-Bond® resin is then applied using two brush techniques. In the case of a large area to cover, the technique of bulk mixing can be used, whereby the monomer and polymer are premixed together in a single vessel and the cement picked up and applied with the brush. For a quicker setting option, the brush-dip technique is used; this consists of first dipping the brush into the polymer solution, draining it of the excess and using it to pick up and apply the monomer powder. In both cases, the bone surface must first be well dried and free of any signs of hemorrhaging.

When the cement is completely dried, the object carrier is raised slowly so that the guide cannula remains in place and visible on the skull (Fig. 6.10a). An obturator ("dummy" cannula) (Fig. 6.10b) is then inserted into the guide so that the latter does not become plugged during the postoperative recovery period (Fig. 6.10c).

It is useful to protect the guide cannula-obturator assembly with a plastic ring of the appropriate dimensions, which encloses the guide cannula but leaves enough space for the operator to easily manipulate the obturator. This protective collar is slipped around the guide cannula and encircled with a second layer of cement to fix it in place (Figs. 6.10c and 6.11).

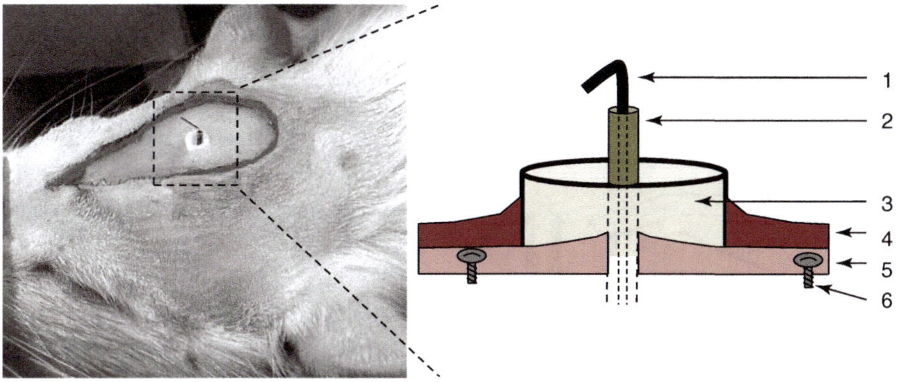

1: Obturator ("dummy" cannula) with angled end for ease of handling
2: Guide cannula
3: Plastic protective ring
4: Second layer of cement
5: First layer of cement
6: Anchor screw

Fig. 6.11 The picture represents detail of the implanted guide cannula enclosed within the cement; the diagram is showing the different components of the material present on the skull surface at the end of the implantation procedure

6.2.2 *Using the Implanted Cannula for Intracerebral Microinjection of Drugs*

For this procedure, the injection cannula is connected to a Hamilton syringe (2, 5, or 10 μl capacity) by a catheter of appropriate dimension; in the example here, the catheter (e.g., polyethylene or Teflon® microtubing) has an internal diameter of 0.3 mm and an outer diameter of 0.7 mm (Fig. 6.12). The syringe and catheter are first filled partially with distilled water. The drug solution is then drawn up through the injection cannula, being careful to leave a small air bubble of about 0.2 μl in the catheter, between the distilled water and the drug solution. Taking the awake animal gently and comfortably but firmly in one hand, the free hand is used to remove the obturator in order to insert the injection cannula. The injection cannula has been prepared with a length greater than that of the guide cannula, so that when inserted its tip will be in direct contact with the target structure.

Once the injection cannula is in place, the drug solution is injected slowly and steadily over a period of 40–60 s. Depending on the target, the injection volume can vary from between 0.1 μl, for a small structure such as the central amygdaloid nucleus, to 4 μl for an intracerebroventricular (i.c.v.) injection. During the injection, the animal must be carefully handled to avoid any sudden movements caused by stress. At the end of the injection procedure, the cannula is left in place for an additional minute or more, to allow for the drug to diffuse away from the site of injection into the tissue of

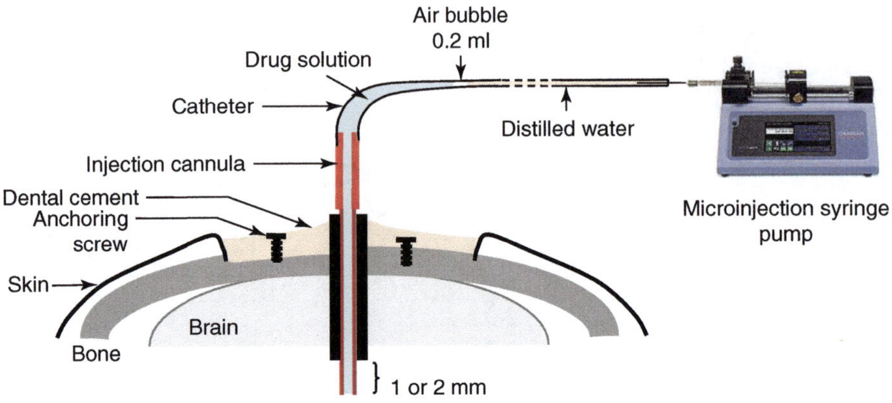

Fig. 6.12 Intracerebral injection device. A microinjection pump is placed on a table along with the injection cannula(e) and their attached catheter(s). The bottom of the guide cannula (*black*) is situated just above the target structure, while the injection cannula (*red*) is 1–2 mm longer to reach the center of the structure after its insertion. Injection cannulae can be kept in 70 % ethanol between each insertion into the cerebral tissue, since their diameter is too small for capillary action to accumulate residual droplets of ethanol (Ferry et al. 1995, 1999, Ferry and McGaugh 2008)

the target structure. This is an important step: if the cannula is withdrawn immediately at the end of injection, the drug solution risks to reflux upward by suction in the wake of the retracting cannula. After the injection cannula has been removed, the obturator is replaced and the animal is then returned to its home cage.

6.2.3 *Using the Implanted Guide Cannula for Activating a Structure by the Optogenetic Method*

Optogenetics is a recent technology based on the use of opsins, a group of transmembrane channel proteins or "channel rhodopsin" that is sensitive to light. Although recent, this technique has already demonstrated its potential: opsins can be incorporated into selected cell types, and their activity can be controlled with a temporal resolution in the order of a millisecond. When one wants to study a structure that is heterogeneous in terms of the neuronal types found within, the optogenetic approach is indispensable in exceeding the limits of specificity imposed by most methods of electrically or neurochemically stimulating or inactivating different neuronal targets. The approach involves five major steps (Fig. 6.13).

The first step consists in elaborating the genetic sequence to be incorporated in the target cells. The gene encoding for opsin is coupled to a promoter expressed in the selected cell population. The second step is to incorporate this sequence into a virus, lentivirus or adenovirus, which provides transfection and proliferation of the sequence. The third step is to stereotaxically inject the viral construction into the target structure.

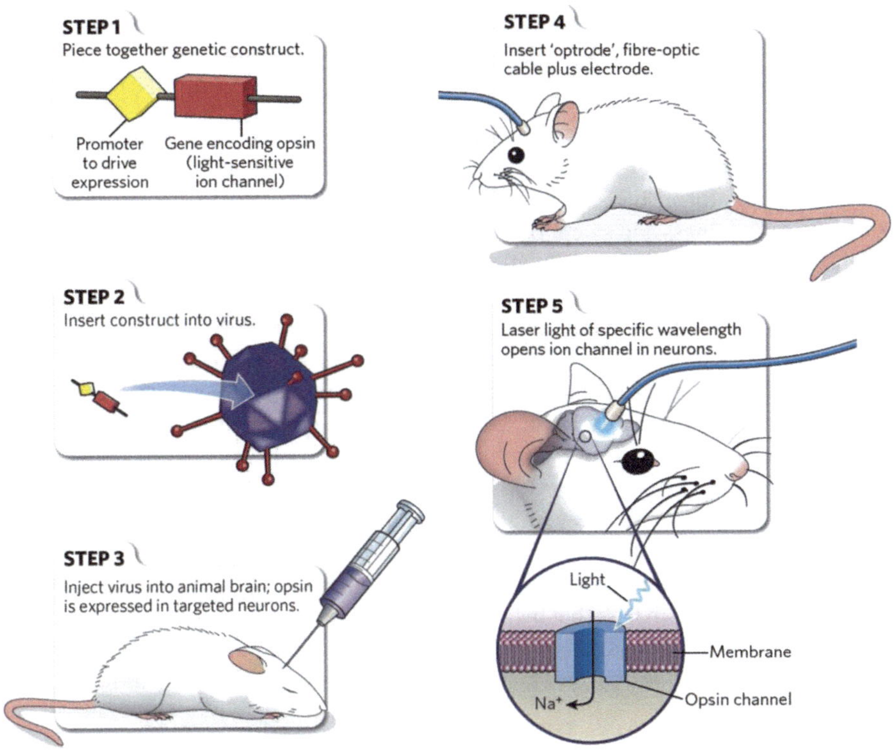

Fig. 6.13 The five steps of the optogenetic technique: the modulation of target neuron activity using light (Adapted from Buchen 2010)

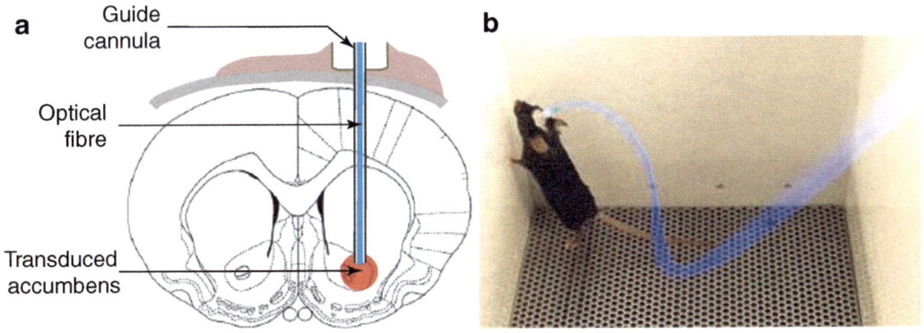

Fig. 6.14 (**a**) Diagram showing a guide cannula and fiber optic for an activation of the nucleus accumbens. (**b**) Conscious animal equipped with the system (Modified from Buchen 2010)

After a delay necessary for the transfection of the targeted neuronal population (about 5–6 weeks minimum), the fourth step is to insert near the structure a fiber optic, coupled (or not) to a recording electrode (Fig. 6.14). After a short delay, the fifth step consists of applying a series of light pulses to obtain a reversible activation (using a blue laser) or an inactivation (using a yellow laser) (Fig. 6.14b).

6.3 Insertion of a Glass Microelectrode in Conscious Rats Under Painless Stereotaxic Head Restraint

First developed in 1994, this technique is useful for carrying out electrophysiological recordings without the interference of general or local anesthetics and, in particular, for monitoring the activity of individual neurons in spontaneously alternating states of wakefulness and sleep. By ensuring a strong and stable mechanical restraint of the animal's head, the device allows for extracellular recordings using delicate and highly sensitive glass micropipettes (Gervasoni et al. 1998, 2000), but it can also be used for the stereotaxic placement of any other type of probe, metallic microelectrodes, microdialysis probes, etc.

Coupled with electroencephalogram (EEG) and electromyogram (EMG) recordings, this preparation can track the spontaneous activity of neurons during the sleep-wake cycle and allow for local microinjections of pharmacological agents or anatomical tracers by iontophoresis or pressure injection. Animals can be used for several daily sessions, in this way serving as their own control and thus allowing a reduction in the number of animals used for the study.

Stereotaxic contention of the head is obtained not by the conventional means of the ear, nose, and tooth bars, but by means of a "U"-shaped piece of metal (Fig. 6.15) permanently fixed to the skull of the animal with dental cement. The attachment is done under general anesthesia. The metal "U" is connected by means of four screws

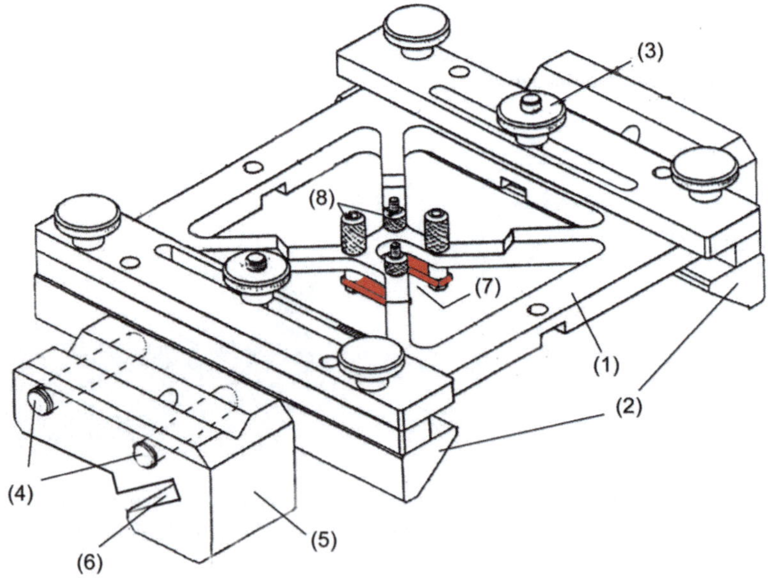

Fig. 6.15 Diagram of the stereotaxic head contention device. The movable carriage (*1*) can slide back and forth on two horizontal transport rails (*2*). The setscrews (*3*) lock the carriage in a given position. Each transport rail has two pins (*4*) engaged in a lateral support bracket (*5*) fixed to the stereotaxic apparatus. During implantation, the ear bars are inserted into the support (*6*), and the metal "U" (*7*, shown in *red*) is attached to the skull of the rat with four screws and dental cement (*8*). The assembly provides a mechanically strong and stable restraint of the animal's head

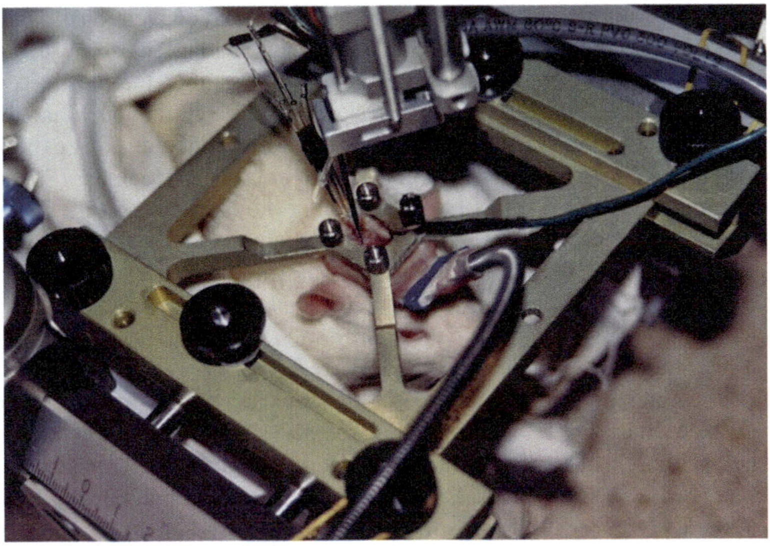

Fig. 6.16 Conscious rat under stereotaxic head contention. The animal, previously habituated to the head restraint, is installed on a hammock suspended on the stereotaxic apparatus. The head is fixed to the center of the device in a position identical to that of the implantation. In this image, the opening of the "U"-shaped piece of metal is facing rearward, to allow a field of access for targeting structures in the pons and medulla. Trepanation of the skull is made for lowering a recording micropipette that is attached to a multichannel pipette for iontophoretic injection of pharmacological agents into the vicinity of the recorded neuron. A microconnector and its attachment cable, visible at the front of the central device, acquire polygraphic signals (EEG, EMG)

to the center of a movable carriage mounted on slides. Two lateral brackets allow mounting the carriage on a conventional stereotaxic instrument.

Since this method of head contention no longer requires the ear or nose bar, it is painless. However, its correct use requires habituation of the animal by progressive training sessions over 10–12 days. After being conditioned with positive reinforcement at the end of each session, the animal is able to spend several hours of contention with normal alternating states of wakefulness and sleep (Fig. 6.16).

6.3.1 *Installation of the Stereotaxic Head Restrainer Device*

Induction of general anesthesia is performed by intramuscular injection of a mixture of ketamine and xylazine (50–90 and 5–10 mg/kg, respectively). During the procedure, anesthesia is maintained by intramuscular injection of half the dose of the same mixture of ketamine and xylazine (25–45 and 5 mg/kg, respectively), at a frequency depending on the level of anesthesia; this is determined by observing the animal's respiratory rate, body temperature, and withdrawal reflex in response to a paw or tail pinch. The duration of anesthesia is approximately 1.5–2 h.

Once surgical anesthesia is obtained, rats are mounted conventionally in a small animal stereotaxic apparatus (e.g., Kopf Instruments) using the ear, nose, and incisor bars. Body temperature is maintained at 37–38 °C using a thermostated heating blanket or pad and a rectal temperature probe. After a longitudinal median incision, the skin overlying the skull is retracted and the surface of the skull exposed and cleaned. The bone is then covered with a thin layer of acrylic cement (Super-Bond C & B®, Sun Medical Co. Ltd., Japan), avoiding the lambdoid suture and an area 5 mm in diameter on the bone above the structure of interest. The metal "U," designed to painlessly immobilize the head of the animal in the stereotaxic apparatus, is placed on its support and its position adjusted above the skull. A well of dental cement (Dentsply De Trey®) is then formed to secure the contention device and the microconnector of the EEG and EMG cables.

After the cement is completely dried, the four screws holding the metal "U" piece to the carriage are removed. Finally, the access well is filled with a dressing (Spongel®, Houde) soaked in antibiotic solution (Neomycin®, Sigma) and covered with bone wax (Horsley®, Braun Surgical). Some sulfanilamide powder (Sigma) is applied to the periphery of the prosthesis to prevent wound infection. After implantation, the animal carries on its head a device of approximately 5 g that allows it to be fixed in a stereotaxic apparatus, with a positioning of the head strictly identical to that of implantation.

6.3.2 Habituation of the Animal to Restraint

After the operation, analgesia is maintained by oral administration of paracetamol (120–200 mg/kg) or Rimadyl®. The nails of the paws are also trimmed to reduce injury from scratching the vicinity of the implant. The general condition of all implanted animals should be monitored daily (Appendices 1 and 2B).

Habituation of animals to head restraint begins 7 days after the surgical procedure and lasts 10–12 days, following a protocol of positive reinforcement conditioning described by Gervasoni et al. (1998, 2000). Wrapped in a loose sock and installed on a hammock, the animals learn to remain with their head fixed in the stereotaxic apparatus for increasingly longer periods. During the training sessions, as soon as the animal begins to show any signs of persistent discomfort, excessive restlessness, vocalization, urination, defecation, or gnashing of the incisors, it is removed from the device. Habituation training ends when the animals are able to stay in contention for 5 h.

6.3.3 Insertion of a Glass Microelectrode

The day before the first recording, a trepanation of 4 mm in diameter is performed under general anesthesia, 4 mm posterior and 1 mm lateral to lambda. The dura

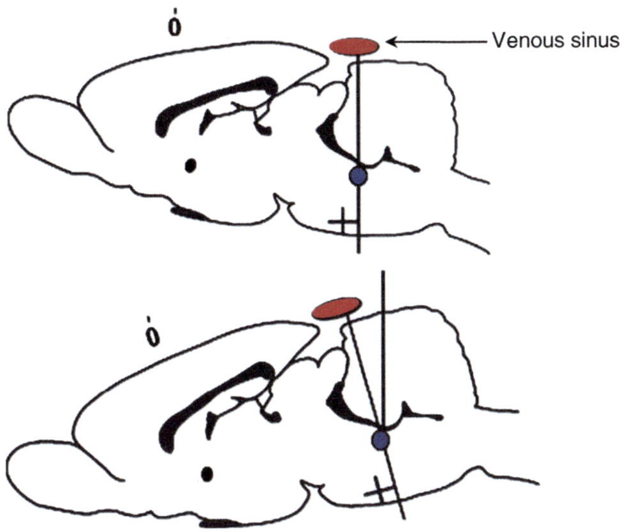

Fig. 6.17 Diagrams in sagittal view showing the position of the transverse venous sinus (in *red*) and the position of the head of the animal after a 15° forward rotation to avoid the blood vessel. The *gray circle* represents the position of the target (the *locus coeruleus*)

mater is exposed, perforated or resected over a small surface area using a sterile beveled needle (see Fig. 6.3b). A new wound dressing is applied and the well is closed as before.

To target a structure such as the nucleus *locus coeruleus* or the dorsal raphe nucleus, which are close to the midline axis, it is necessary to use an angle due to the presence of the transverse venous sinus. The obstacle that this blood vessel presents to the glass micropipette can be bypassed by a forward inclination of the head at an angle of around 15° or by an inclination of the electrode holder (Fig. 6.17).

In both cases, new coordinates are recalculated. In the following example, the head of the animal is tilted rather than the electrode holder, in order to maintain the verticality of the latter. When working at an angle, bear in mind that any displacement of the microelectrode or probe in the brain takes place simultaneously in two directions, in this case dorsoventral and anteroposterior, the point of entry at the brain surface being caudal relative to the deeper target point.

Extracellular action potentials are recorded by means of glass microelectrodes, made from drawn glass micropipettes, tips broken to have an external diameter of 3 μm and filled with a solution of sodium acetate (0.5 M) containing 2 % w/v of pontamine blue dye (Fig. 6.18). The reference electrode (zero potential) is made from a silver chloride wire place in a well filled with a solution of 0.9 % sodium chloride. The recording electrode is lowered vertically into the brain in successive stages or "steps" with the aid of a hydraulic micro-stepping motor (Micropositioner®, Kopf Instruments). Extracellular action potentials are recorded simultaneously with the electroencephalogram (EEG) and electromyogram (EMG) for monitoring and recording the state of vigilance of the animal. At the end of the study, the site of implantation is verified, as described in Fig. 7.11 (see Chap. 7).

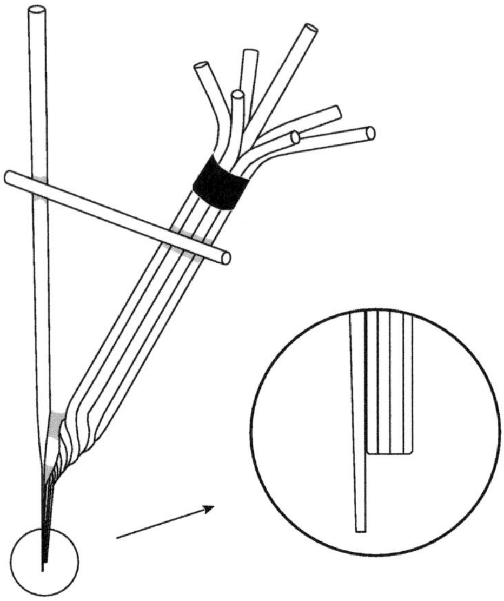

Fig. 6.18 Assembly of a recording electrode and a multichannel pipette. The multichannel pipette is mounted "piggy-back" configuration 15–20 μm back from the tip of the recording microelectrode (Urbain et al. 2002, 2006)

6.4 Intracerebral Microdialysis in the Anesthetized Mouse

Once the mouse is anesthetized (Messier et al. 1998), it is placed in a stereotaxic apparatus, in this example one that incorporates a Cunningham mouse adaptor (Figs. 6.19 and 6.20).

The location of the bregma and lambda reference points is similar to the rat (Fig. 6.21). However, the suture lines are generally difficult to observe; it is common to apply a solution of 3 % hydrogen peroxide in order to improve the contrast.

After verification of the horizontality of the skull, the stereotaxic coordinates are used to locate the point of insertion for the microdialysis probe.

In this case the coordinates are the following: AP, – 1.8; ML, 1.5; and DV, – 2.4 from the dura mater.

The dorsoventral coordinate should be taken when the probe is positioned on the skull just above the target structure. The skull of the mouse, like that of the rat, is domed; from this coordinate, the probe will be finally inserted (Fig. 6.22).

For insertion of a microdialysis probe (Fig. 6.23), the method is identical to that used for injection cannulae or electrodes, the only adaptations necessary being the amount of dental cement and the size of the cylinder used to protect the upper part of the probe.

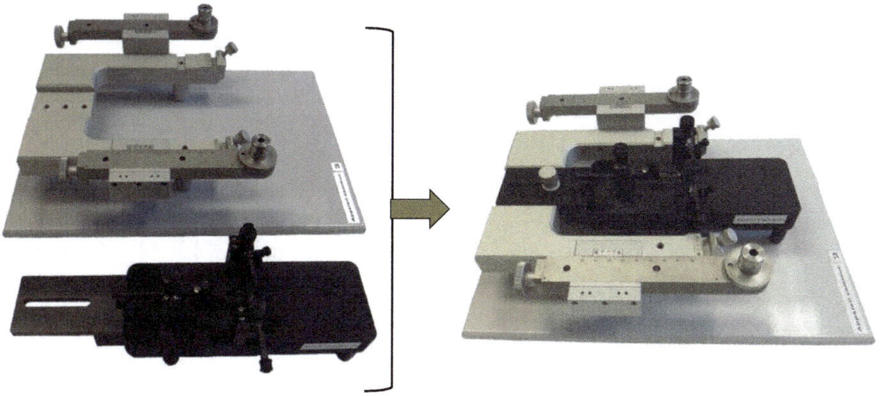

Fig. 6.19 Photographs of a Cunningham stereotaxic adaptor (*black* device for mice), installed in a conventional small animal stereotaxic instrument frame

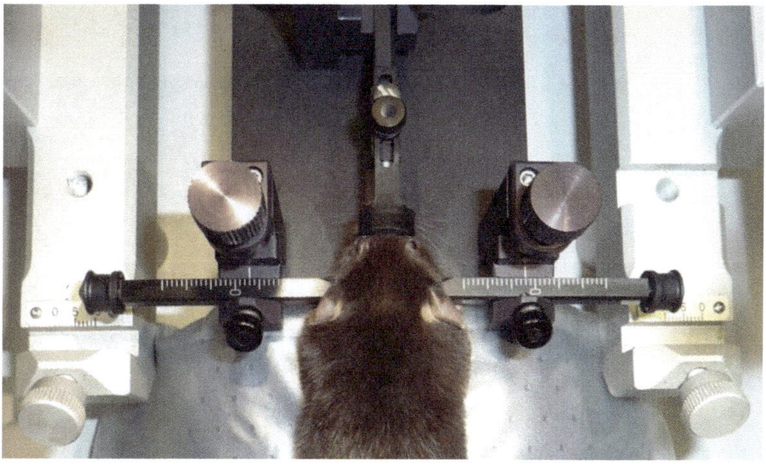

Fig. 6.20 The procedures for fixing the head of the mouse with the nose and ear bars of the Cunningham adaptor are essentially identical to those for the rat

Difficulties posed by targets situated ventrally to a major vascular structure are treated according to the same principles as in the rat (see Sects. 4.2.2 and 4.2.4 and Fig. 6.17).

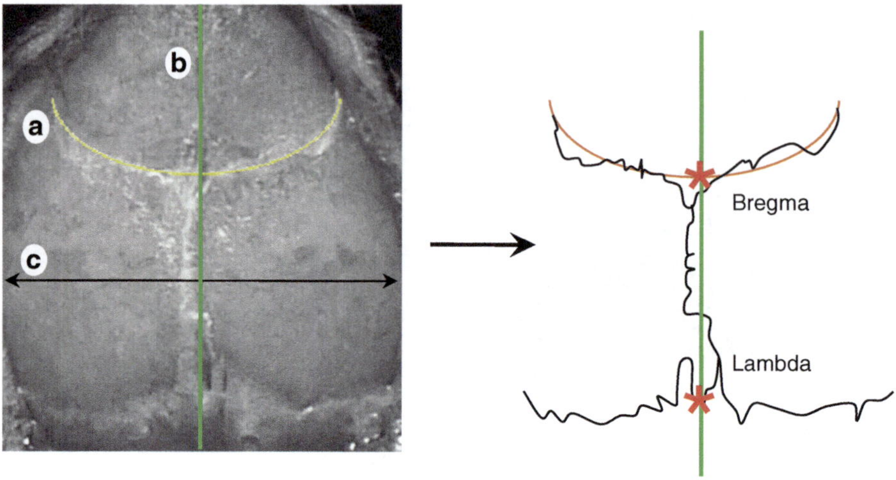

Fig. 6.21 Photograph of the surface of a mouse skull showing the bregma and lambda reference points (Messier et al. 1998; Blasiak et al. 2010)

Fig. 6.22 Placement of a microdialysis probe in mice. In this example, the probe will perfuse the dorsal hippocampus

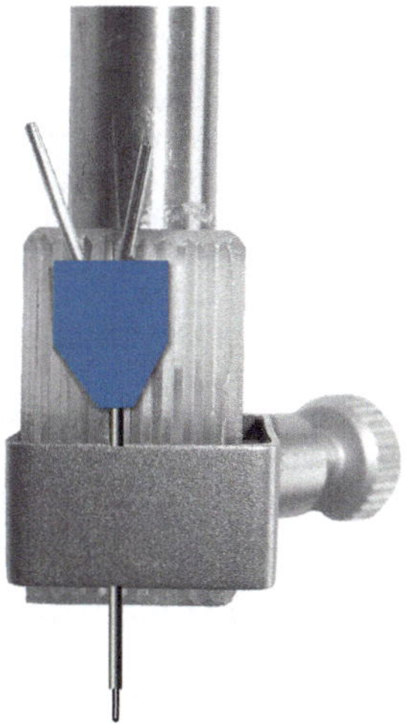

Fig. 6.23 Placement of a microdialysis probe on the object carrier. Before proceeding with insertion of the probe into the brain, the verticality of the axis of descent must be verified

References

Berridge KC, Whishaw IQ (1992) Cortex, striatum and cerebellum: control of serial order in a grooming sequence. Exp Brain Res 90(2):275–290

Blasiak T, Czubak W, Ignaciak A, Lewandowski MH (2010) A new approach to detection of the bregma point on the rat skull. J Neurosci Methods 185(2):199–203

Blum D, Torch S, Lambeng N et al (2001) Molecular pathways involved in the neurotoxicity of 6-OHDA, dopamine and MPTP: contribution to the apoptotic theory in Parkinson's disease. Prog Neurobiol 65(2):135–172

Buchen L (2010) Illuminating the brain. Nature 465:26–28

Centonze D, Calabresi P, Giacomini P, Bernardi G (1999) Neurophysiology of Parkinson's disease: from basic research to clinical correlates. Clin Neurophysiol 110(12):2006–2013

Chen Z, Ong BH, Kambouris NG, Marban E, Tomaselli GF, Balser JR (2000) Lidocaine induces a slow inactivated state in rat skeletal muscle sodium channels. J Physiol 524:37–49

Dudley MW, Howard BD, Cho AK (1990) The interaction of the beta-haloethyl benzylamines, xylamine, and DSP-4 with catecholaminergic neurons. Annu Rev Pharmacol Toxicol 30:387–403

Duncan RJS, Sourkes TL, Dubrovsky BO, Quik M (1975) Activity of aldehyde dehydrogenase, aldehyde reductase and acetylcholine esterase in striatum of rats bearing electrolytic lesions of the medial forebrain bundle. J Neurochem 24(1):143–147

Ferry B, McGaugh JL (2008) Involvement of the basolateral amygdala α_2-adrenoceptors in memory modulation of inhibitory avoidance in the rat. Learn Mem 15(4):238–243

Ferry B, Sandner G, Di Scala G (1995) Neuroanatomical and functional specificity of basolateral amygdaloid nucleus in taste potentiated odor aversion. Neurobiol Learn Mem 64:169–180

Ferry B, Wirth S, Di Scala G (1999) Functional interaction between entorhinal cortex and basolateral amygdala during trace conditioning of odor aversion in the rat. Behav Neurosci 113(1):118–125

Fritschy JM, Grzanna R (1991) Selective effects of DSP-4 on locus coeruleus axons: are there pharmacologically different types of noradrenergic axons in the central nervous system? Prog Brain Res 88:257–268

Gervasoni D, Darracq L, Fort P, Souliere S, Chouvet G, Luppi PH (1998) Electrophysiological evidence that noradrenergic neurons of the locus coeruleus are tonically inhibited by GABA during sleep. Eur J Neurosci 10:964–970

Gervasoni D, Peyron C, Rampon C et al (2000) Role and origin of the GABAergic innervations of dorsal raphe serotonergic neurones. J Neurosci 20:4217–4225
Glinka Y, Gassen M, Youdim MB (1997) Mechanism of 6-hydroxydopamine neurotoxicity. J Neural Transm 50:55–66
Goldstein LB (1993) Beam-walking in rats: the measurement of motor recovery after injury to the cerebral cortex. Neurosci Protoc 10(3):1–13
Jarrard LE (2002) Use of excitotoxins to lesion the hippocampus: update. Hippocampus 12(3):405–414
Jonsson G (1980) Chemical neurotoxins as denervation tools in neurobiology. Annu Rev Neurosci 3:169–187
King BM (1991) Comparison of electrolytic and radio-frequency and lesion methods. Methods Neurosci 7:90–97
Kostrzewa RM (1995) Dopamine receptor supersensitivity. Neurosci Biobehav Rev 19(1):1–17
Kostrzewa RM (2009) Evolution of neurotoxins: from research modalities to clinical realities. Curr Protoc Neurosci 1: Unit 1.18
Kostrzewa RM, Brus R (1998) Destruction of catecholamine-containing neurons by 6-hydroxydopa, an endogenous amine oxidase cofactor. Amino Acids 14(1–3):175–179
Li Y, Peris J, Zhong L, Derendorf H (2006) Microdialysis as a tool in local pharmacodynamics. AAPS J 8(2):222–235
Livingston-Thomas JM, Hume AW, Doucette TA, Tasker RA (2013) A novel approach to induction and rehabilitation of deficits in forelimb function in a rat model of ischemic stroke. Acta Pharmacol Sin 34(1):104–112
McGaughy J, Everitt BJ, Robbins TW, Sarter M (2000) The role of cortical cholinergic afferent projections in cognition: impact of new selective immunotoxins. Behav Brain Res 115(2):251–263
Messier C, Emond S, Ethier K (1998) New techniques in stereotaxic surgery and anesthesia in the mouse. Pharmacol Biochem Behav 63(2):313–318
Milner TA, Amaral DG (1984) Evidence for a ventral septal projection to the hippocampal formation of the rat. Exp Brain Res 55(3):579–585
Rotman A, Creveling CR (1976) A rationale for the design of cell-specific toxic agents: the mechanism of action of 6-hydroxydopamine. FEBS Lett 72(2):227–230
Schliebs R, Rossner S, Bigl V (1996) Immunolesion by 192IgG-saporin of rat basal forebrain cholinergic system: a useful tool to produce cortical cholinergic dysfunction. Prog Brain Res 109:253–264
Schwarting RK, Huston JP (1996) The unilateral 6-hydroxydopamine lesion model in behavioral brain research. Analysis of functional deficits, recovery and treatments. Prog Neurobiol 50(2–3):275–331
Schwarting RK, Huston JP (1997) Behavioral and neurochemical dynamics of neurotoxic mesostriatal dopamine lesions. Neurotoxicology 18(3):689–708
Urbain N, Rentero N, Gervasoni D, Renaud B, Chouvet G (2002) The switch of subthalamic neurons from an irregular to a bursting pattern does not solely depend on their GABAergic inputs in the anesthetic-free rat. J Neurosci 22:8665–8675
Urbain N, Creamer K, Debonnel G (2006) Electrophysiological diversity of the dorsal raphe cells across the sleep-wake cycle of the rat. J Physiol 573:679–695
Vargo JM, Marshall JF (1996) Unilateral frontal cortex ablation producing neglect causes time-dependent changes in striatal glutamate receptors. Behav Brain Res 77(1–2):189–199
Wiley RG (1992) Neural lesioning with ribosome-inactivating proteins: suicide transport and immunolesioning. Trends Neurosci 15(8):285–290
Wiley RG (1997) Findings about the cholinergic basal forebrain using immunotoxin to the nerve growth factor receptor. Ann NY Acad Sci 835:20–29
Wiley RG (2001) Targeting toxins to neural antigens and receptors. Methods Mol Biol 166:267–276

Chapter 7
Post-mortem Examination of Animals

At the end of a whole animal experiment, postmortem examination is necessary to validate the experimental data. Steps include euthanasia, autopsy and removal of the brain, treatments specific to organ fixation, sectioning, and staining.

7.1 Different Types of Euthanasia Accepted by Legislation (Directive 2010/63)

The term "euthanasia" is derived from the Greek *eu* (good) and *thanatos* (death) and literally means a "good death," ethically speaking.

Accordingly, methods of humane euthanasia must comply with a number of principles. While being compatible with the experiment, the method chosen must also bring about death without pain or signs of panic or stress. Loss of consciousness must be quick or instantaneous. The technique must be reliable and reproducible but also proven safe for the animal handler. Euthanasia must be performed in a place separate from premises where the animals are normally housed.

The method of euthanasia most commonly used in neuroscience is anesthetic overdose. This consists of administering several times the dose of an anesthetic, such that a rapid loss of consciousness is induced and death by cardiopulmonary arrest is provoked.

Cervical dislocation consists of provoking injury of the cervical spinal cord and rupture of the blood vessels supplying the brain. Death is rapid and without pharmacological interference (from anesthetics, analgesics, etc.). However, the technique requires experience and is usually reserved for mice and rats smaller than 200 g body weight.

Decapitation meets similar criteria as cervical dislocation. The procedure consists of cutting off the head of the animal at its base, using the appropriate instrument: a very sharp and strong knife blade or guillotine, which should be thoroughly cleaned after each use, so as not to stress subsequent animals by the smell of blood. The gesture of execution must be confident and without hesitation.

B. Ferry et al., *Stereotaxic Neurosurgery in Laboratory Rodent*,
DOI 10.1007/978-2-8178-0472-9_7,

It is always preferable to carry out the above two procedures after the animals have already been anesthetized.

Euthanasia with carbon dioxide is the method of choice for rodents, especially when carrying out the procedure on numerous experimental groups. Loss of consciousness is rapid and death occurs within 2–3 min.

Exsanguination is used only under deep anesthesia. This method allows for the collection of a large volume of blood sometimes necessary for certain experimental measurements and at the same time permits preparation of the animal for organ fixation by perfusion.

> *Note*: After euthanasia, it is imperative to verify the death of the animal.

7.2 Perfusion

Fixation of the brain by perfusion is usually performed under deep anesthesia of the animal while the heart is still beating. The procedure requires the use of the appropriate instruments (Fig. 7.1) to access the thoracic cavity and to put the perfusion catheters in place.

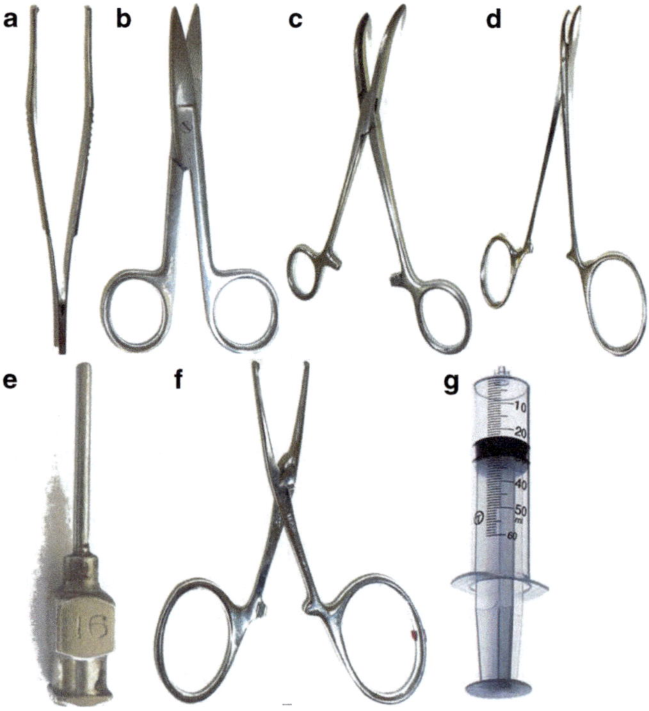

Fig. 7.1 Instruments used for the perfusion and removal of the brain. (**a**) Toothed tissue dissecting forceps. (**b**) Mayo operating scissors. (**c**) Curved hemostatic forceps. (**d**) Needle holder. (**e**) Round-tipped (blunt) syringe needle. (**f**) Straight hemostatic forceps. (**g**) 50 ml syringe

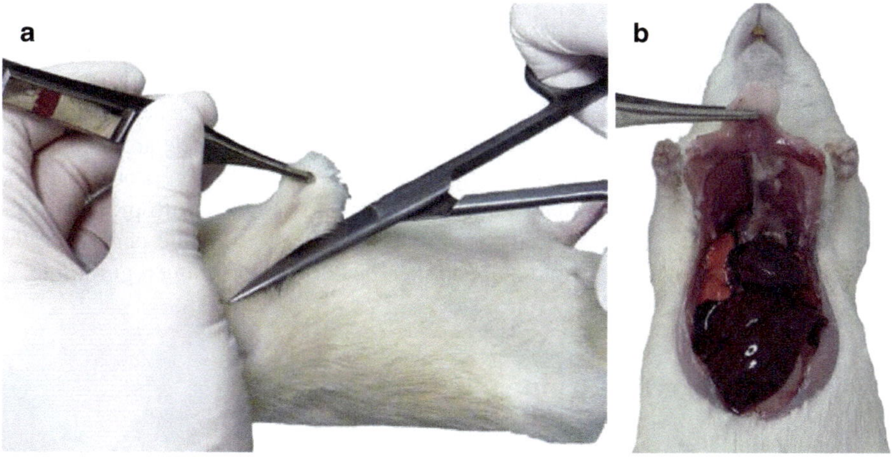

Fig. 7.2 (**a**) Gripping the sternum with forceps before the incision with operating scissors. (**b**) Retraction of the rib cage towards the head of the animal after lateral incisions of the diaphragm

Once deep anesthesia has been induced, the thorax is opened. Using a hemostatic forceps (Fig. 7.1c), the xiphoid process (the inferior part of the sternum) is gripped through the skin. By applying a slight upward traction, the costal arch (or costal margin) can be visualized under the skin. An incision with scissors is made obliquely through the skin, muscle, and the ribcage on both sides of the sternum (Fig. 7.2a). The diaphragm is freed at its lateral insertions in order to retract the anterior wall of the thorax towards the head of the animal (Fig. 7.2b).

This surgical approach gives wide access to all of the thoracic organs, especially the heart. The heart is retracted to the left in order to see the thoracic aorta, which is clamped entirely without prior dissection. In this way, the posterior part of the animal is isolated from a vascular point of view, leaving the volume of the intracardiac perfusion to distribute only within the anterior part. The perfusion needle (see Fig. 7.1e) is inserted into the left ventricular muscle and slipped inside the ventricular cavity up to the start of the aorta. In order not to damage the aorta itself, a blunt round-tipped needle should be used. Once the needle is in place, it is clamped together with the heart with the aid of the second pair of hemostatic forceps (Fig. 7.1f), in order to maintain the needle in place within the heart during the perfusion. The right atrium is partially sectioned with scissors to allow for the venous outflow of the perfusion liquid. The perfusion is initiated with 50 ml of physiological saline to eliminate a maximum amount of blood from the arteriovenous cerebral circulatory system.

Perfusion then continues with 50 ml of a 4 % paraformaldehyde solution to prefix the organ. All of the following procedures should be performed under a vertical laminar flow hood to avoid the risk of exposure to formaldehyde vapors.

With whatever liquid used, the perfusion can be performed manually if it is done slowly enough to minimize intravascular back pressure. However, it is preferable to use a peristaltic pump adjusted to a fixed maximum flow rate of 20 ml/min. A successful perfusion of paraformaldehyde is evident when it reaches the muscles of the forelimbs. At this point, spasmodic movements of the muscle are induced by the conformational change of muscle protein in reaction to the fixative. Formaldehyde is an electrophilic substance that binds covalently upon contact with proteins (rich in nucleophilic sites) causing their denaturation. At the end of the perfusion, the body of the animal becomes rigid and can then be decapitated using a guillotine.

7.3 Removal of the Brain

To ensure an optimum quality of fixation, the brain is removed before the rest of the body is examined. Figure 7.3 illustrates the equipment required. Once the head has been removed using a guillotine, the scalp is incised medially in order to laterally retract the skin and muscle overlying the skull surface. Bone rongeurs are used to remove pieces of the bone plates, starting at the occipital foramen at the base of the skull.

Once the cerebellum has been exposed, the extremity of one of the jaws of the rongeurs is carefully inserted between the dura mater and the brain, taking care not

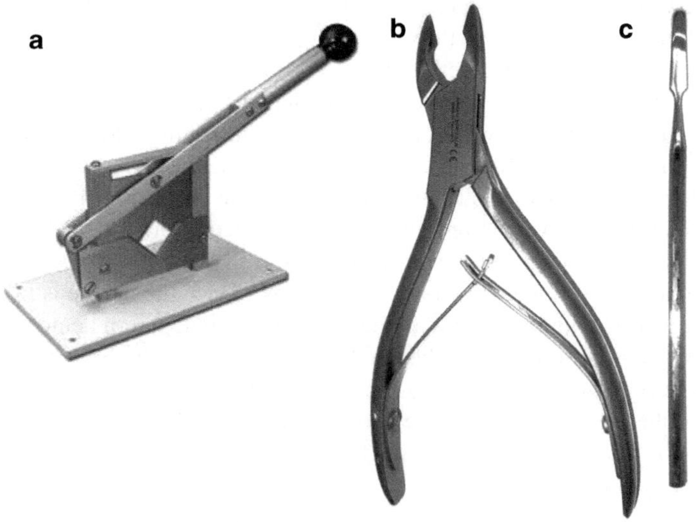

Fig. 7.3 Instruments used for removing the brain. (**a**) Small animal guillotine. (**b**) Bone rongeurs. (**c**) Spatula

to damage the surface of the latter. The various plates of cranial bone are removed one by one by carefully lifting upward from the inside out. When the overlying dura and bone are completely disengaged from the brain, a spatula (see Fig. 7.3j) is used to detach and lift the brain off of the floor of the cranial vault. The brain is then immersed in a fixation solution of 4 % paraformaldehyde at 4° C for 2 h, followed by a 36 h immersion in 30 % sucrose before carrying out the cryoprotection procedure.

7.4 Autopsy and Examination of the Organs in the Large Body Cavities

At the end of the perfusion, the large (thoracic and abdominal) cavities of the body should be opened to make a rapid assessment of the size, shape, superficial aspect, and color of the various internal organs, in order to eventually reveal any abnormalities in these features (Fig. 7.4) (Piper et al. 2011). The severity of certain observations can serve as a preliminary signal of disease, malformation, or toxicity to the researcher and the attendant veterinarian. If the lesions are associated with a disease state, a diagnosis may be possible through complementary examinations by a diligent practitioner. Depending on the circumstances, it may also be informative to the animal supplier or breeder of a problem or anomaly.

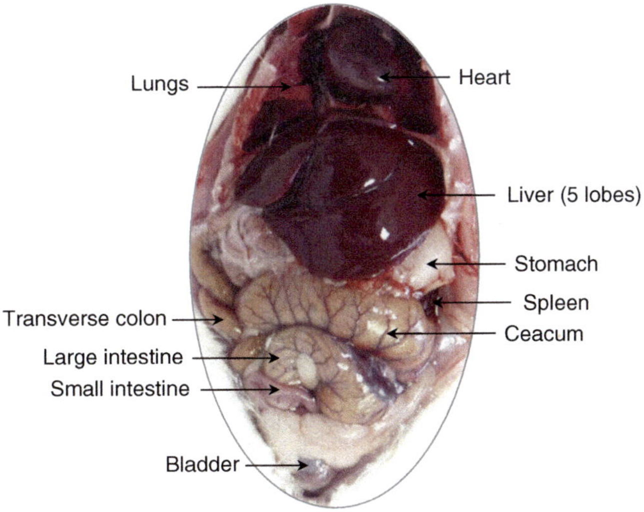

Fig. 7.4 View of the internal organs after median incision of the thorax and abdomen of the rat

7.5 Preparation of the Brain for Visual Examination on a Cryostat

To obtain high-quality brain sections, a maximum of air contained within the brain tissue must be removed if not eliminated during the freezing process. Fixed brains should be cooled as quickly as possible. It is advisable to start by immersion in isopentane previously cooled to −40° C with dry ice for about 1 min. To ensure stringent temperature conditions, a thermometer should be used to monitor the temperature and the isopentane should be stirred just before the brain is immersed therein. Once frozen, the brain is removed from the solution using smooth forceps or a spatula and then securely fixed on the cutting stage of the cryostat using an appropriate adhesive matrix made for this purpose.

When the experimenter wants to verify and possibly modify any deviations of the stereotaxic coordinates, one or more implantation sites can be done in "pilot" animals. After implantation of one or two injection cannulae, the animal is euthanized. Injection of 0.1–0.2 µl of a tracer, such as bromothymol blue (Fig. 7.5), neutral red (Fig. 7.6), or cresyl violet (Fig. 7.7) is made through the cannula, in this way marking the actual zone that was targeted by the coordinates used for the cannula implantation. Macroscopic observation of serial frozen sections at the level of the marked zone allows in as few as one animal and at the very start of the research project a validation of and the

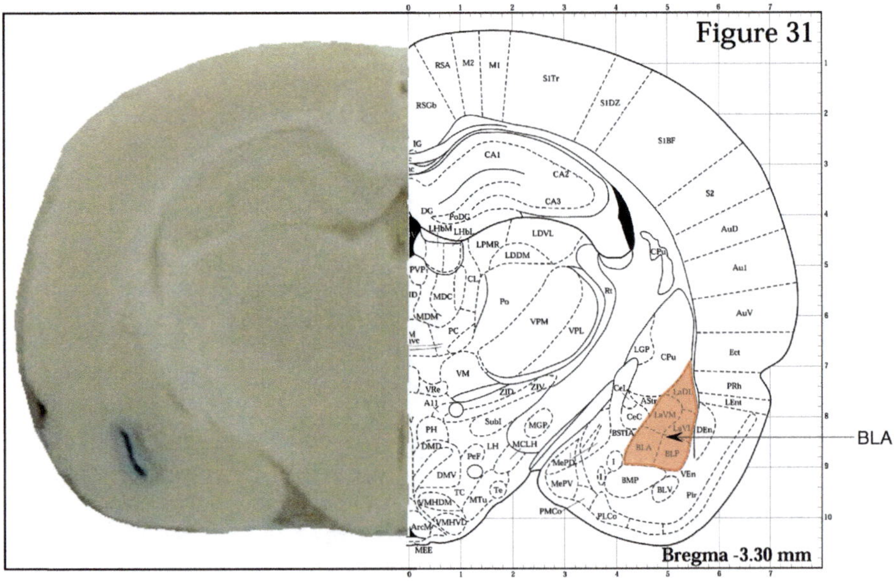

Fig. 7.5 Verification of coordinates for the basolateral nucleus of the amygdala (BLA) in the rat using bromothymol blue. *Left*: photo of the rat brain mounted in the cryostat after being sectioned up to the point where the *blue-colored* tracer can be discerned. *Right*: diagram of the corresponding plane of section from the atlas of Paxinos and Watson (2004)

possibility to correct, if needed, the stereotaxic coordinates of the target structure. This is especially useful when the target region is very small.

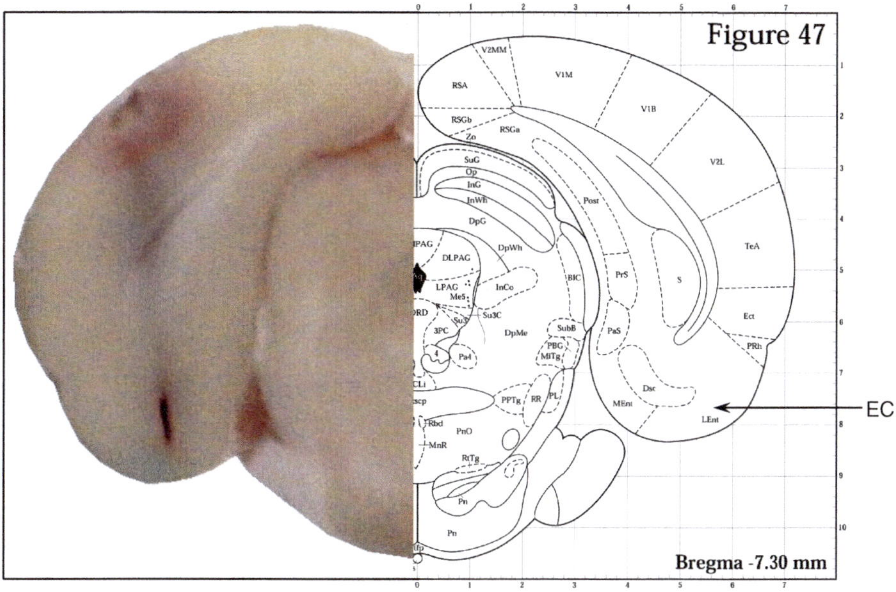

Fig. 7.6 Verification of coordinates for the entorhinal cortex (EC) in the rat using neutral red. *Left*: photo of the rat brain mounted in the cryostat after being sectioned up to the point where the *red-colored* tracer can be discerned. *Right*: diagram of the corresponding plane of section from the atlas of Paxinos and Watson (2004)

7.6 Preparation of Brain Sections for Histology

It is possible to begin sectioning of brains immediately after the freezing procedure. Sections of frozen brain are made using a cryostat at a temperature between −19 and–23° C. Section thickness can be varied between 10 and 30 μm depending on the purpose of the study. When there are a large number of brains to section, they can be temporarily stored within the cryostat chamber so that they will be at the same temperature as the cryostat blade when sectioning begins. Brain sections can be mounted on cold gelatinized glass slides, but mounting is much easier with slides previously exposed to room temperature. Once the cold brain section makes contact with the warmer slide, it naturally tends to spread itself out flat; if not, it can be carefully unfolded or unrolled with the aid of a fine camel hairbrush moistened with distilled water.

Brain sectioning is performed on all animals used in the study. At the end of the sectioning procedure, slides are dried by placing them on a rack and storing them at room temperature for 24 h or placing them in an oven at 37 °C for 12 h. After staining, e.g., with cresyl violet (Appendix 10), each section is examined under an optical microscope. It is helpful to have on hand copies of the relevant

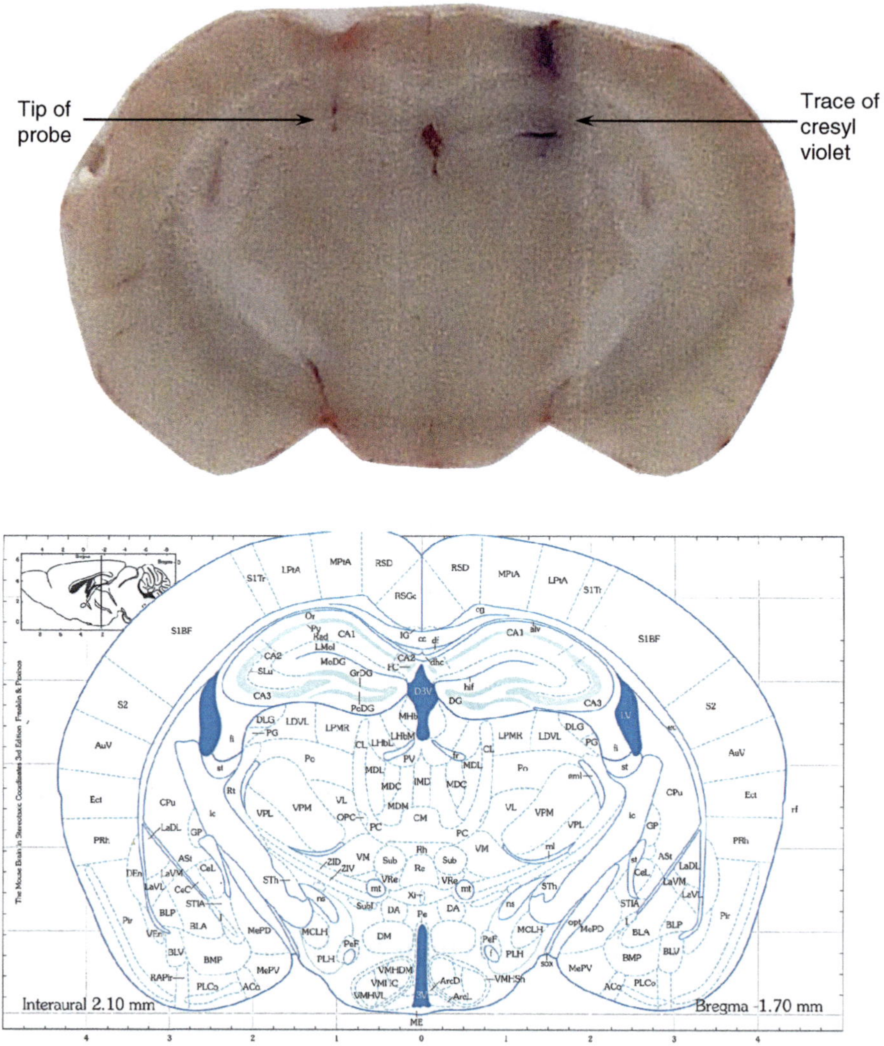

Fig. 7.7 Verification of coordinates for the dorsal hippocampus in the mouse brain using cresyl violet. *Top*: photo of a brain from an animal used in a microdialysis experiment. In the right hemisphere (*left side* of photo), the tract of the microdialysis probe and its point of termination can be seen. In the contralateral hemisphere (*right side* of photo), the trace of a cresyl violet injection can be seen; this was used as a reference for eventual adjustment of the final placement coordinates of the probe (at a point 0.1 mm higher) in subsequent microdialysis experiments. *Right*: diagram of the corresponding plane of section from the mouse brain atlas of Franklin and Paxinos (2008)

plates taken from the stereotaxic atlas, upon which one can directly note the animal's identification and record the actual location of each site of implantation.

Following a review of all the sections, only those animals that have had the correctly targeted location validated are retained for statistical analyses. This histological control is the final step of stereotaxic surgery and is a critical phase for the experimenter. Providing evidence for the site of manipulation (Stevens et al. 2002; Young et al. 2006, Young and Heath 2000) permits a verification of the coordinates under which the experiments were designed and performed. It is

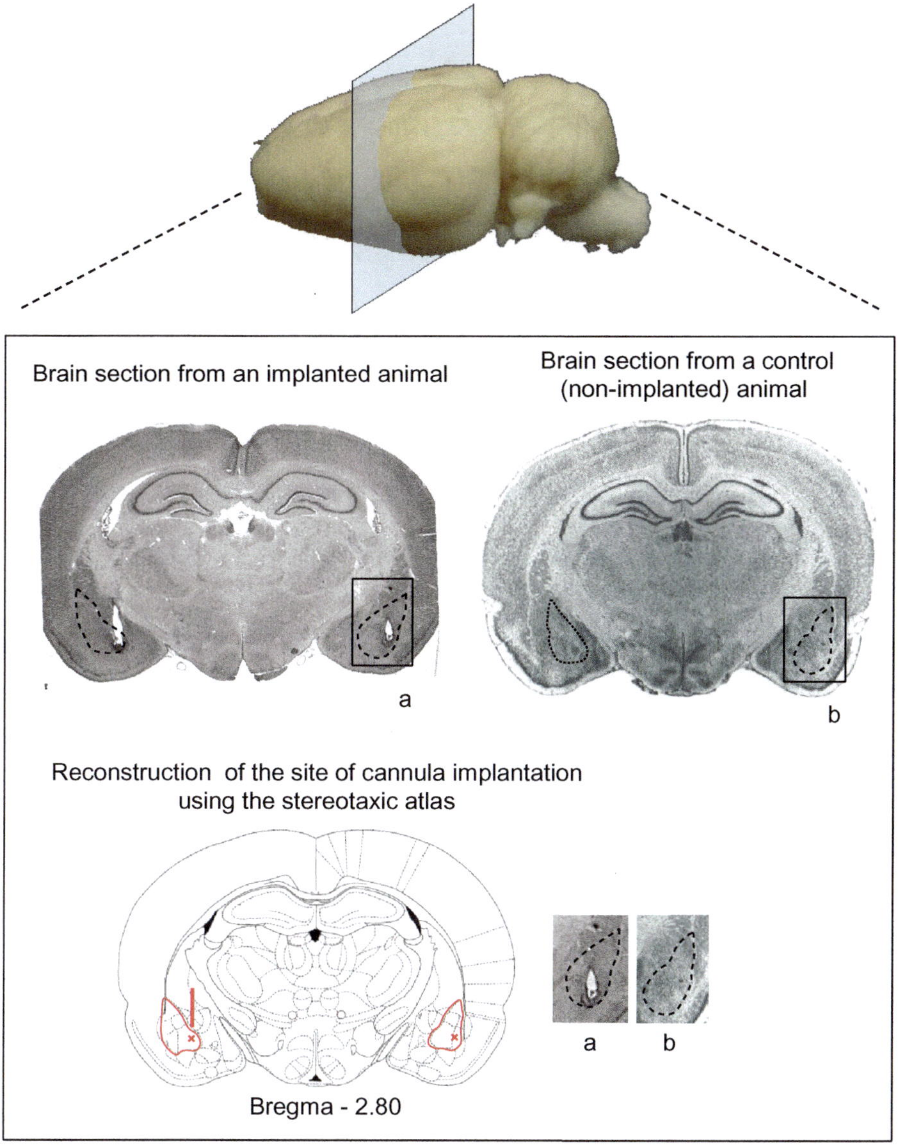

Fig. 7.8 Frontal sections of the rat brain (*upper images*) showing the implantation site of a guide and injection cannula (Ferry et al. 1995, Ferry and McGaugh 2008). Using a stereotaxic atlas, the localization of the injection site is determined by examining a series of adjacent brain sections. The diagram in the *lower left* of the figure is a plate from the atlas of Paxinos and Watson (2004), corresponding to the level (2.8 mm posterior to bregma) of the brain sections shown in the upper images. (**a**) Enlarged view of the site of implantation in the basolateral nucleus of the amygdala. (**b**) Enlarged view of the same target site in a control (non-implanted) animal

only after this verification that correlations between acquired data and target site can be made (Ross and Pawlina 2006).

- Example 1: determining the site of implantation of a guide and injection cannulae in the basolateral nucleus of the amygdala (Fig. 7.8)

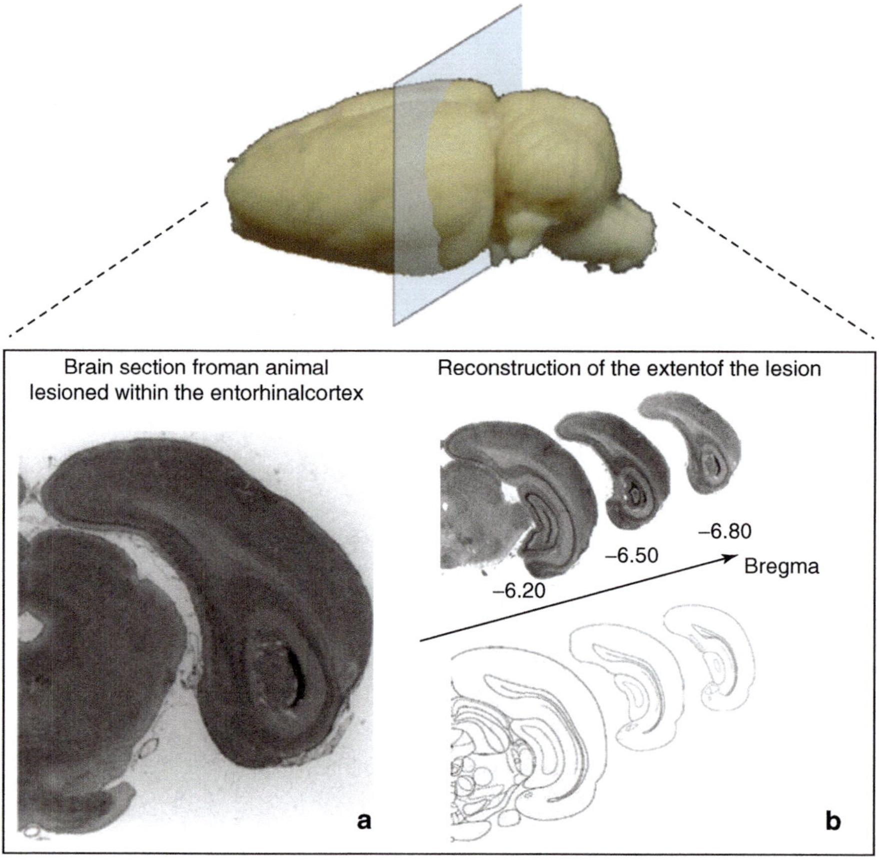

Fig. 7.9 Frontal brain sections showing the site of an experimentally induced lesion within the entorhinal cortex of the rat (Ferry et al 1996, 2006). With the aid of the stereotaxic atlas and the examination of a series of adjacent sections to determine the anterior and posterior limits, the full extent of the lesion can be assessed. (**a**) Frontal section of the left hemisphere of an animal with a lesion in the entorhinal cortex. (**b**) Frontal sections of the left hemisphere showing the lesion within the entorhinal cortex at three different levels, from −6.2 to −6.8 mm posterior to bregma, according to the stereotaxic atlas of Paxinos and Watson (2004). Representation of the extent of the lesion is shown as the grayed zones in the *lower diagram*

- Example 2: determining the extent of a lesion the entorhinal cortex using frontal sections (Fig. 7.9)
- Example 3: determining the extent of a lesion in the entorhinal cortex using horizontal sections (Fig. 7.10)
- Example 4: determining the site of implantation of a glass microelectrode in the locus coeruleus of the rat (Fig. 7.11)

Basic histological procedures can be employed for different labeling techniques depending on the purpose of the experiment. Thus, in the relatively

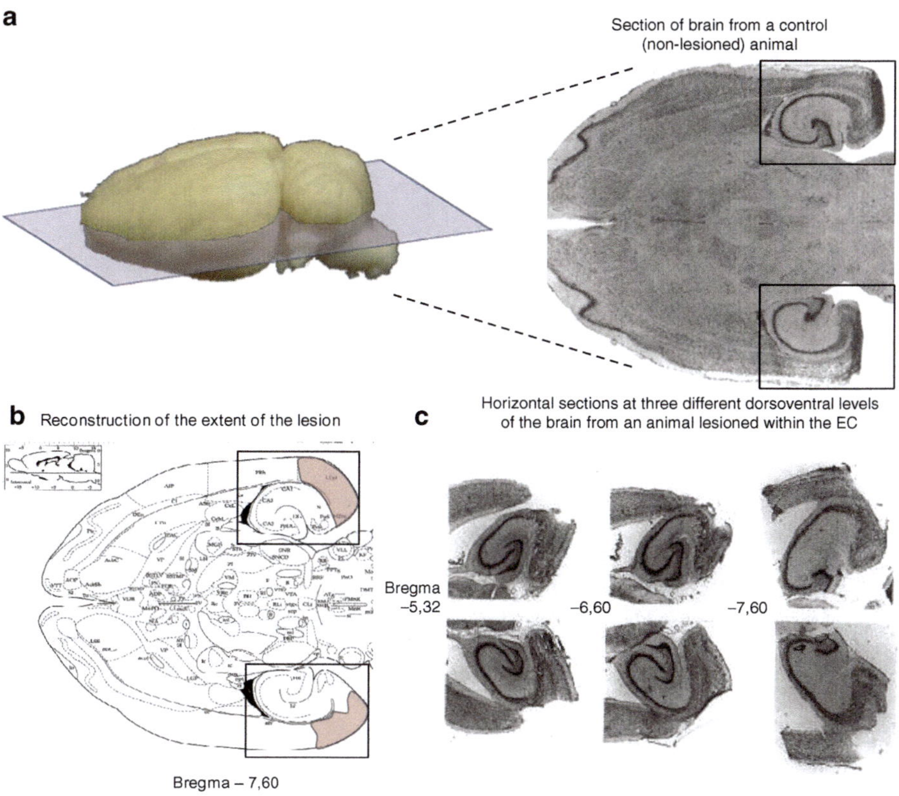

Fig. 7.10 Horizontal sections of the brain showing the site of an experimentally induced lesion within the entorhinal cortex (EC) of the rat. The stereotaxic atlas is used here to assess the full extent of the lesion by determining its start and endpoints in a series of adjacent sections. (**a**) Horizontal section from a non-lesioned control animal. (**b**) Plate from the atlas of Paxinos and Watson (2004) corresponding to the horizontal plane at 7.6 mm ventral to bregma. The extent of the lesion observed in the brain sections shown in (**c**) is represented by the *red-shaded areas*. (**c**) Horizontal sections of brain showing both hemispheres of an animal lesioned in the entorhinal cortex (Ferry et al. 1999, 2006)

simple case of visualizing the tract of a cannula or a microelectrode or the localization of an experimentally induced brain lesion, a simple labeling of brain structures using conventional stains such as methylene blue and cresyl violet can be sufficient. However, for more sophisticated labeling techniques such as acetylcholinesterase staining, in situ hybridization, or immunohistochemistry, tissue preparation protocols *specific* to each of these different labeling methods need to be used. Since the accuracy of targeting the experimental site of interest conditions the exploitation of the results obtained and their eventual publication, it is paramount to proceed with these types of verification with a maximum of precision.

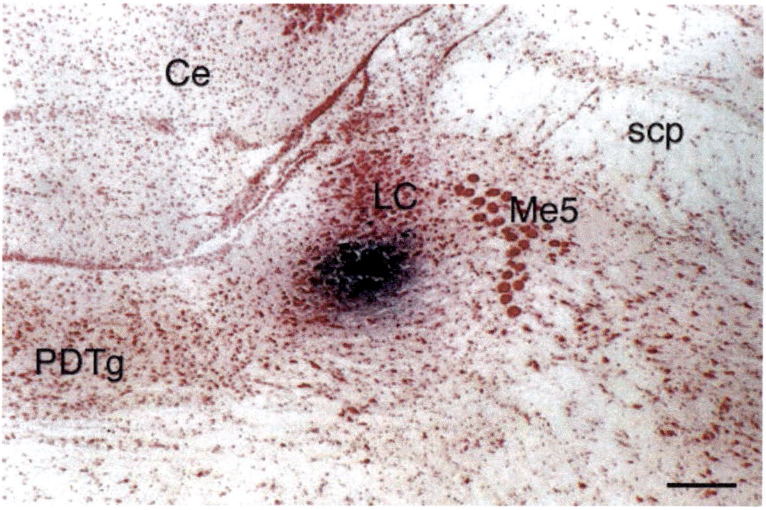

Fig. 7.11 Frontal section of rat brain showing the location of a microelectrode recording site within the locus coeruleus (LC). Pontamine sky blue was deposited in situ by microiontophoresis at the end of the recording experiment (Gervasoni et al. 1998, 2000). The staining is visible on frontal sections of unfixed brain mounted on glass slides and counterstained with neutral red

References

Directive 2010/63/EU du Parlement Européen et du Conseil (22 septembre 2010) sur la protection des animaux en expérimentation scientifique. Journal officiel de l'Union Européenne (Official Journal of the European Union), Strasbourg

Ferry B, Sandner G, Di Scala G (1995) Neuroanatomical and functional specificity of basolateral amygdaloid nucleus in taste potentiated odor aversion. Neurobiol Learn Mem 64:169–180

Ferry B, Oberling P, Jarrard LE, Di Scala G (1996) Facilitation of conditioned odor aversion learning by entorhinal cortex lesions in the rat. Behav Neurosci 110(3):443–450

Ferry B, Wirth S, Di Scala G (1999) Functional interaction between entorhinal cortex and basolateral amygdala during trace conditioning of odor aversion in the rat. Behav Neurosci 113(1):118–125

Ferry B, Ferreira G, Traissard N, Majchrzak M (2006) Selective involvement of the lateral entorhinal cortex in the control of the olfactory memory trace during conditioned odor aversion in the rat. Behav Neurosci 120(5):1180–1186

Ferry B, McGaugh JL (2008) Involvement of the basolateral amygdala α_2-adrenoceptors in memory modulation of inhibitory avoidance in the rat. Learn Mem 15(4):238–243

Franklin KBJ, Paxinos G (2008) The mouse brain in stereotaxic coordinates, 3rd edn. Oxford Academic Press, Oxford

Gervasoni D, Darracq L, Fort P, Souliere S, Chouvet G, Luppi PH (1998) Electrophysiological evidence that noradrenergic neurons of the locus coeruleus are tonically inhibited by GABA during sleep. Eur J Neurosci 10:964–970

Gervasoni D, Peyron C, Rampon C et al (2000) Role and origin of the GABAergic innervations of dorsal raphe serotonergic neurones. J Neurosci 20:4217–4225

Paxinos G, Watson C (2004) The rat brain in stereotaxic coordinates, 5th edn. Academic, San Diego

Piper M, Treuting S, Dintzis M (2011) Comparative anatomy and histology: a mouse and human atlas. Elsevier Academic Press, San Diego: USA

Ross MH, Pawlina W (2006) Histology: a text and atlas with correlated cells and microbiology, 5th edn. Lippincott Williams and Wilkins, Philadelphia, pp 318–352

Stevens A, Lowe J, Young B (2002) Wheater's basic histopathology: a text and color atlas, 4th edn. Churchill Livingstone, Edinburgh

Young B, Heath JW (2000) Wheater's Functional Histology, 4th edn. Churchill Livingstone, Edinburgh

Young B, Lowe J, Stevens A, Heath, JW (2006) Wheather's functional histology: a text and color atlas, 5th edn. Churchill Livingstone, Edinburgh

Appendices

Appendix 1: Reporting Abnormalities in Experimental Animals

Report Checklist

- Describe the species, strain, age, weight, and sex.
- Describe any abnormal clinical signs: size, position, coloration, etc.
- Isolate the cage.
- Assess the animal's environment.
- Check the feeder and water bottle (or watering system).
- Check the inside of the cage and the litter for abnormal fecal material or urine.
- Observe the behavior and attitude of the animal in its home cage.
- Open the cage and proceed with the examination of the animal.
- Note the animal's general appearance, fur and color of the skin, any abnormal movements, thinness or obesity, discharges or excessive secretions, etc.
- Note the reaction of the animal when handled (aggressiveness, hypotonia, dehydration).
- Examine the animal's back, abdomen, and body extremities.
- Check for any misalignment or overgrowth of incisors.
- Check the animal's records for any previous clinical observations, history, surgery, etc.

Be familiar with the characteristics of the particular species and strain being examined, since important differences do exist (alopecia, obesity). In any case, consult with the principal investigator and/or veterinarian in charge.

B. Ferry et al., *Stereotaxic Neurosurgery in Laboratory Rodent*,
DOI 10.1007/978-2-8178-0472-9, © Springer-Verlag France 2014

Appendix 2A: Typical Follow-Up Record of Stereotaxic Surgery

Name of operator :	Protocol or project number / code :
Animal id number / code :	Species / strain / sex :
Study / experimental number :	N° cage :
Lot number :	Date of reception :
Room number :	Date of surgery :

Type of surgical operation: Pre-operatory body weight:

Anesthetic		Date and time of administration	Analgesic		Date and time of administration
Type	*Dose*		*Type*	*Dose*	

Record of stereotaxic coordinates:

Target area / region :			
Atlas coordinates:	AP: - Y	ML: ± X	DV: - z
Coordinates used :	AP	ML	DV
Bregma	A	B	C
Lambda	U	B	C
Calculation of target coordinates	A - Y	B ± X	V - Y
Connector or cannula (e;g. 1 mm above the target)			V – Y - 1
Strereotaxic apparatus station / n°			
A, B, C, U *are arbitraryunits used here as an example*; V is the DV coordinate measured above the target			

Visual localization:

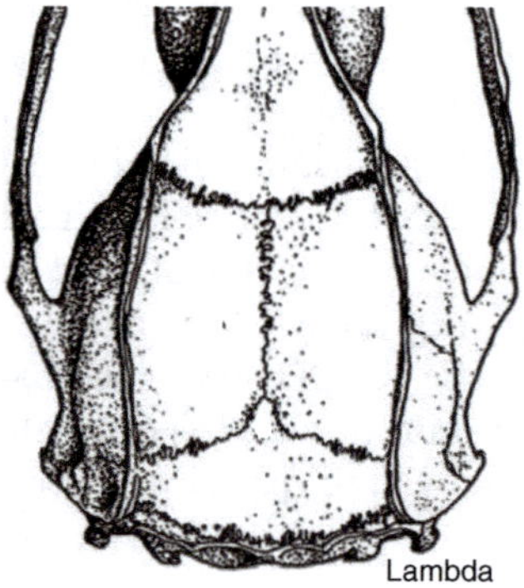

This surgical follow-up form should be retained by the research investigator in his/her laboratory notebook. A part of the form should be reserved for use by animal care staff to record any additional treatments that may eventually be carried out.

Appendix 2B: Typical Postsurgical Follow-Up Report with Daily Observations

Animal identification number :

Monitoring of body weight (in grams), e.g. 300 g on the day of surgery

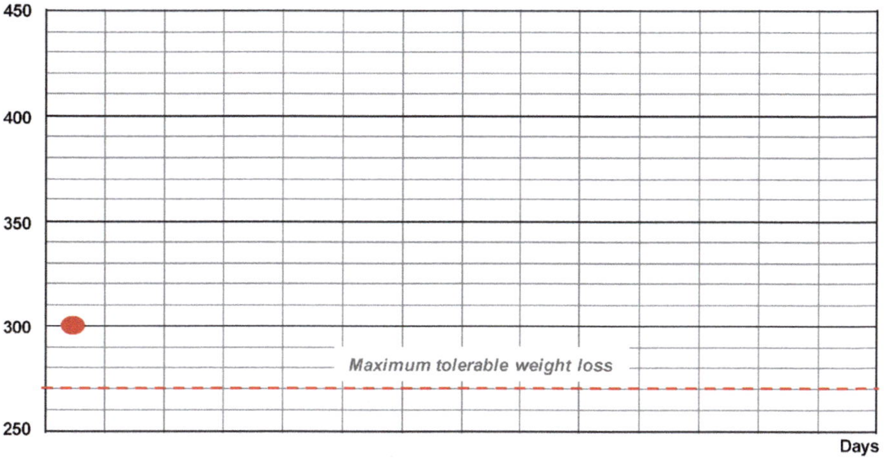

Date	Behavior	Food / Drinking water	Signs of pain, discomfort	Treatment	Observations

Remarks :

Any animal euthanized or found dead should be retained for autopsy by the principal investigator or veterinarian.

Appendix 3: General Criteria for Deciding on Euthanasia

During the course of an experiment, an endpoint is reached when, considering the objectives of the study, any pain and/or distress of the animal needs to be minimized, reduced, or stopped. Measures must be taken to terminate any procedure that produces suffering, either by appropriate interventions and treatments of the animal or by its humane euthanasia. In practice, the decision to euthanize requires the use of an evaluation tool for assessing pain or discomfort, based on different but easily observable criteria.

Examples of Criteria

- Anorexia and weight loss
- Self-injurious behavior, alopecia (loss of hair), and skin abrasions
- Altered state of consciousness: lethargy or apathy
- Modified, irregular, difficult, or noisy breathing; coughing, rattling, wheezing, or whistling
- Abnormal color/appearance of the mucous membranes: anemia, jaundice, or congestion
- Development of spontaneous tumors affecting the animal's normal behavior and habits
- Abnormal neurological signs: inclination of the head, tremors, spasticity, convulsions, stereotypies, paresis, paraplegia, and paralysis
- Changes in locomotor function affecting drinking and food intake

These criteria are evaluated quantitatively and regularly, in order to score the condition of the animal relative to a preestablished and quantifiable threshold score defined by consensus as the endpoint. All available means (analgesic or other) are used to help the animal regain a state of health and behavioral well-being that will allow its comfort and survival. However, if the threshold is reached or exceeded, the animal must be euthanized.

Appendix 4: Taking Care of Pain: From Observation to Decision to Execution

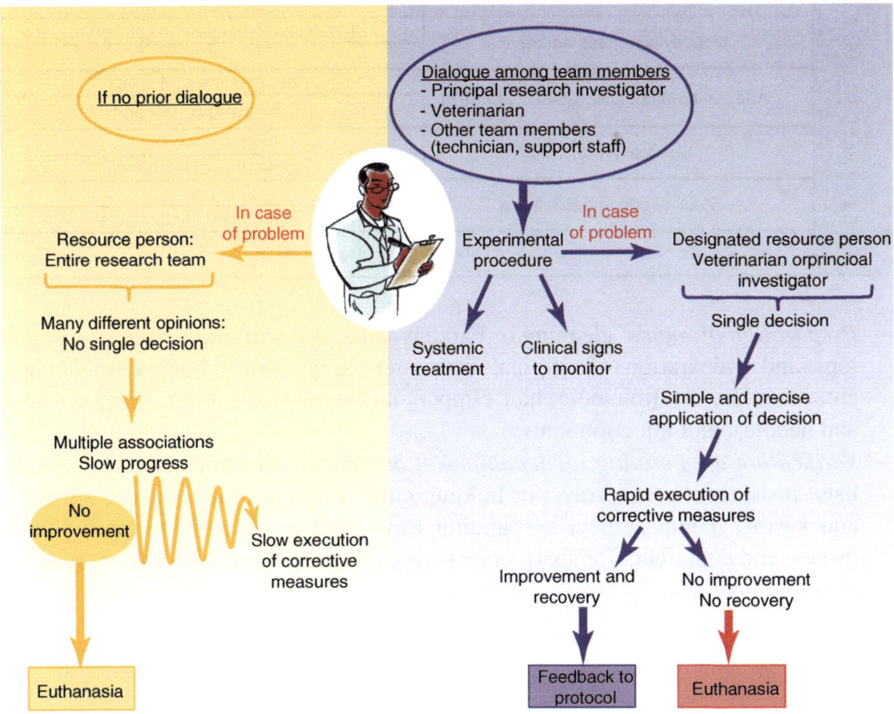

To ensure and facilitate a proper postoperative follow-up, a simple decisional tree should be established at the outset of the experimental protocol, through dialogue between active and accessory participants in the study. This dialogue will help to define a procedure that addresses any eventual clinical signs to be aware of, the systemic treatments to use if necessary, and the contact person (principal investigator and/or veterinarian) in case of problems. This flowchart is intended to provide a quick and single unambiguous decision and a simple and precise action plan of corrective measures. If problems are common, it may be necessary to modify the experimental protocol, provide operators with additional training in surgical or animal care procedures, or increase the postoperative recovery time of the animal.

Appendix 5: Checklist for Surgical Procedure Setup

THE DAY BEFORE SURGERY	Stereotaxic base and frame	Verniers
		Electrode holders
	Magnifying lenses and optics	Light bulbs, other sources of illumination
		optics
	Artificial respirator, anesthetic vaporizer	Anesthetics fluid levels
		Tubing and connections
	Surgical intruments	Instrument boxes
		Sterilization of instruments
	Stock of consumable items	Drugs
		Sutures and compresses
	Verification of working order and cleanliness of equipment required for the procedure	

- *Preparation of rooms:* cleaning of Faraday cages, magnifying lenses, and bench tops and preparation of surgical instrument trays, wash bottles of alcohol, chlorhexidine or Hibitane®, hair clippers and combs, anesthesia tray, syringes and needles, and injection vials.
- *Verification of operating room equipment and materials*: instrument tray checklists; instrument cases, trays, or holding cups; surgical drapes, sterile sponges, and swabs; dental cement preparation trays; and organic waste bins, kidney dishes, and containers for sharp objects (e.g., disposable scalpel blades, syringe, or suture needles).

The Day of Operation

Phase	Step	Details	
1H before	Clinical examination of the animal	Observation of behavior	
		Weigh-in	
NON-STERILE	SURGICAL GLOVES AND GROWNS		
	Analgesia and anesthesia	Injection of Rimadyl®(NSAI)	
		Sedation	
		Induction of anesthesia	
	Disinfection of the paws, tail and operatory field	Shaving	
		Cleaning of the surgical zone (chlorhexidine SCRUB (Hibitane®, Iodine antiseptics Betadine®)	
		Alcohol wash	
	Anesthesia	Injection or intubation	
	Positioning the animal	Heating blanket or pad, temperature probe	
		Ear bars, nose clamp, incisor bar	
STERILE	SURGICAL GLOVES AND GOWNS		
	Placing the sterile drapes	Opening of drape box	Surgical assistant
	Surgical instrument tray	Opening of surgical instrument boxes, blades, compresses	
		Fill disinfectant cups	
	Aseptic operating zone	Fill irrigation cups	
	Placement of waterproof surgical barrier drape		
	Incision		
	Dissection		
	Hemostasis		
	Hydrogen peroxide wash	Disposal of used compresses	
	Checking the positioning of the animal	Respecting the chain of asepsis	
	Taking coordinates of the stereotaxic reference points		
	Drilling		
	PROTECTION OF SURGICAL WOUND WITH COMPRESSES + STERILE PHYSIOLOGICAL SALINE SOLUTION		
NON-STERILE	Technical procedure: verification of electrodes, connectors, surgical screws, dental cement, adhesive resin...		
	DISINFECTION WITH CHLORHEXIDINE SPRAY		
	CHANGE OF SURGICAL GLOVES		
STERILE	ACUTE ELECTROPHYSIOLOGY	CHRONIC ELECTROPHYSIOLOGY	ACUTE MICRODIALYSIS / CHRONIC MICRODIALYSIS

RECOVERY FROM ANESTHESIA

The animal is maintained on the heating blanket, and then placed in heated holding cage or box until it awakens.

or

EUTHANASIA		Dissection of tissue samples for histological examinations
Deep anesthesia	Perfusion	

OPERATING BLOC CLEAN-UP	**KEEP GLOVES ON, PUT ON SURGICAL GOWN**	
	Gather up contaminated items	Organic wastes
		Sharp object waste containers
	Management of surgical equipment	Soak surgical instruments and related equipment
		Drain briefly, wipe and let dry
	CLEAN AND DISINFECT BENCHES, STEREOTAXIC STAND AND OPERATING MICROSCOPE USING CHLOREXIDINE SPRAY OR TOWELETTES	
	Disposal of wastes	Collect reusable surgical linens for washing
		Empty the waste bins
	REMOVE GLOVES AND GOWN THEN CLOSE THE OPERATORY BLOCK	

THE DAY AFTER THE OPERATION	Arrange the clean dry instruments in their boxes
	Close the box with sterilization indicator tape, record date and operator name, sterilize for 1.5H at 180°C
	Refill the compresses boxes
	Package the boxes of surgical linens
	close the box with sterilization indicator tape, record date and operator name, sterilize for 1.5H at 180°C
	Restock the instrument trays
	Check the stock of injectables (anesthetics, analgesics)
	Store away clean equipment and materials

Appendix 6: Ethics in Animal Experimentation

Since ancient times, medical practitioners have used animals to satisfy their intellectual curiosity about structural and functional anatomy. It was not until the early seventeenth century, however, that British physiologists began to be concerned about the welfare of experimental animals. Since then, considerable efforts have been made to replace animals with alternative models for teaching the life sciences, for example, by using cell cultures derived from skin or tumors, lower vertebrates (bacteria, yeast, shellfish, and insects), or computer simulations. Over time these alternative techniques have allowed a reduction in the number of animals used in research and teaching. These methods have their limitations; however, the complexity of an entire intact living organizing cannot be reproduced in the laboratory by a simple layer of single uniform cells, by bacteria, or by a computer model. Thus, when it can be shown that there is no alternative method available to achieve the objectives of research having an established scientific or medical need, recourse is given to the ethical use of animals. Ethics define and clarify the concepts that characterize good or bad practices and recommend corrective measures accordingly. Whenever the use of animals in research is discussed, ethical issues and animal rights must obligatorily be taken into account.

The Principles of the "3Rs"

After the First World War, certain groups interested in the welfare of animals used for scientific purposes came into being, including the Universities Federation for Animal Welfare (UFAW). The UFAW commissioned William Russell, a philosopher, and Rex Burch, a microbiologist, to write a manual, *The Principles of Humane Experimental Technique (1959)*. This seminal work first introduced the notion of the "3Rs," which became a focus point for more than 2000 scientists, veterinarians, animal care technicians, students, community representatives, and animal welfare associations who contributed to the ethical review system and supervision of the Canadian Council on Animal Care (CCAC) established in Canada since 1968. These three principles are "replacement," "reduction," and "refinement."

Replacement often means the use of an inanimate system as an alternative (e.g., mannequins, computer models or programs, multimedia teaching aids such as videos). It can also mean the replacement of sentient animals (usually vertebrates) with less sentient animals (usually invertebrates such as worms and bacteria). It also includes the use of cell and tissue cultures. The cells must come from somewhere and often this means animals.

Reduction means a decrease in the number of animals used previously with no loss of useful information. This can be achieved by reducing the number of variables through good experimental design, by using genetically homogeneous animals, or by ensuring that the conditions of the experiment are rigorously controlled.

When animals cannot be replaced, their number should be limited to the minimum number required to achieve statistically significant results. Often this can be determined a priori by a statistical "power analysis" to estimate the requisite sample size (minimum number of animals per experimental group). Protocols should not unnecessarily repeat previously validated studies.

Refinement means a change or improvement in some aspect of the experiment that results in a reduction or replacement of animals or in a reduction of any pain, stress, or distress that animals may experience. The establishment of early endpoints for intervention in a study that has the potential to cause pain or distress is an example of refinement. The use and development of techniques and noninvasive technologies can contribute to refinement, improving the quality of experimental results by minimizing the discomfort of the animal (better procedures in anesthesia and analgesia, magnetic resonance imaging, telemetry for remote recording of temperature, blood pressure and heart rate, etc.).

- A fourth "R" has been defined by the University of Sherbrooke (Canada) as the "responsibility" of the individual investigator with respect to the use of animals for research purposes, i.e., credibility and scientific justification for using experimental animals as well as empathy, compassion, sensitivity, and attention to their physical and psychological well-being.

Appendix 7: Nonsteroidal and Opioid Analgesics and Dosages

Molecule	Rat	Mouse (hamster, gerbil)
Paracetamol (acetaminophen)	200 mg/kg p.o. 24 h	120 mg/kg p.o. 24 h
Carprofen	5 mg/kg IM, SC or p.o. 24 h	5 mg/kg SC 12 h
Flunixin	2.5 mg/kg IM or SC 12 h	2.5 mg/kg IM or SC 12 h
Ketoprofen	2.5 mg/kg IM, SC or p.o. 24 h	
Meloxicam	1–2 mg/kg SC or p.o. 12 h	2.5 mg/kg IM, SC or p.o. 24 h
Buprenorphine	0.01–0.05 mg/kg SC or IV 8 to 12 h 0.1–0.25 mg/kg p.o. 8 to 12 h	0.01–0.05 mg/kg SC 8 to 12 h
Butorphanol	1–2 mg/kg IM or SC 2 to 4 h	1–2 mg/kg IM or SC 2 to 4 h
Morphine	2–5 mg/kg IM or SC 4 h	2–5 mg/kg SC 4 h
Nalbuphine	1–2 mg/kg IM 4 h	2–4 mg/kg IM 4 h
Pentazocine	5–10 mg/kg IM or SC 3 to 4 h	5–10 mg/kg IM or SC 3 to 4 h
Pethidine (meperidine)	10–20 mg/kg IM or SC 2 to 3 h	10–20 mg/kg IM or SC 2 to 3 h
Naloxone antagonist	0.01–1 mg/kg IM, SC or IV	

IM intramuscular, *SC* subcutaneus, *IV* intravenous, *p.o.* per os

Appendix 8: Anesthetics and Dosages

	MOUSE	RAT	DURATION ANESTHESIA / RECOVERY TIME	ANTAGONISTS OR RESPIRATORY STIMULMANTS	
Ketamine (Imalgène®, kétamine-Virbac, Ketanest, Ketaset, Ketalar) + Acepromazine (acetylpromazine; Calmivet, Vétranquil)	100 mg/kg 2.5 mg/kg IM, IP	75 mg/kg 2.5 mg/kg IM, IP	20 - 30 min / 2 h		Doxapram® (Dopram): 5-10 mg/kg IM, IV or IP respiratory stimulant, effective for all types of anesthesia including inhalational
Ketamine (Imalgène®) + Diazepam (Valium®)	100 mg/kg 5 mg/kg IP	75 mg/kg 5 mg/kg IP	20 min / 2 h		
Ketamine (Imalgène®) + Xylazine (Rompun®)	50 to 90 mg/kg + 5 to10 mg/kg IM, IP	90 mg/kg + 10 mg/kg IM, IP	20 - 30 min / 2 to 4 h	Yohimbine, non specific antagonist 0.2 mg/kg IV ou 0.5 mg/kg ; not recommended IM	
Ketamine (Imalgène®) + Medetomidine (Domitor®)	75 mg/kg + 0.5 mg/kg IP	75 mg/kg + 0.5 mg/kg IP	20 - 30 min / 2 to 4 h	Atipamezole, specific antagonist 1 mg/Kg IM, IP, SC ou IV	
Urethane	-	1000 mg/kg IP	6 - 8 h / WITHOUT RECOVERY		
Metomidate (Marinil, Methoxynol) + Fentanyl (Sublimaze, Actiq, Duragesic, Fentora…)	60 mg/kg 0.06 mg/kg SC	-	30 min / 1 h	Reverse opioid-induced respiratory depression and analgesia Naloxone (Narcan, Nalone, Narcanti): 0.01 – 0.1 mg/kg IV, IP or IM	
Fentanyl + Fluanison (Hypnorm®)	0.01 ml/30g IP	0.4 ml/kg IM, IP	20 min / 60 min +		
Fluanisone (Hypnorm®) + Diazepam (Valium®)	0.3 ml/Kg 5 mg/kg IP	0.3 ml/kg 2.5 mg/kg IP	45 - 60 min / 3 - 4 h +	Reverses respiratory depression and maintains analgesia Nalbuphine : 1 - 2 mg/kg/ 3h IM	
Fluanisone (Hypnorm®) + Midazolam (Hypnovel®)	10 ml/kg IP	2.7 ml/kg IP	45 - 60 min		
	Mixture: 1 part fluanisone + 1 part midazolam (5mg/ml) + 2 parts water		3 - 4 h +		

IM intramuscular, *SC* subcutaneus, *IV* intravenous

Appendix 9: Overview of Suturing Materials

While the sterility of suturing materials is a paramount prerequisite, a good tolerance or biocompatibility is of equal importance. The presence of a ligature causes an inflammatory reaction of the body, provoking an inflammatory granuloma that is particularly susceptible to infection. The intensity of the reaction depends upon the material (natural or synthetic) that constitutes the suture thread, its structure (monofilament or braided), and its capillarity. The breakdown process (hydrolysis or proteolysis) and time required for suture resorption are also contributing factors. Resorption time is a decisive factor in the choice of the thread, since the ligature needs to remain effective during the entire healing process. To prevent diffusion of liquids into the spaces between the filaments of the thread, it is preferable to use a non-capillary material such as a monofilament thread. Braided (multifilament) threads that are coated to reduce capillarity and drag can also be used. The strength of the thread must be able to resist tensile forces when the ligature is made, to last until the end of the wound healing process, and, despite a 40–50 % loss of this characteristic at the level of the knot, to allow for doing an additional knot.

Types of Suture Thread

Braided (multifilament)

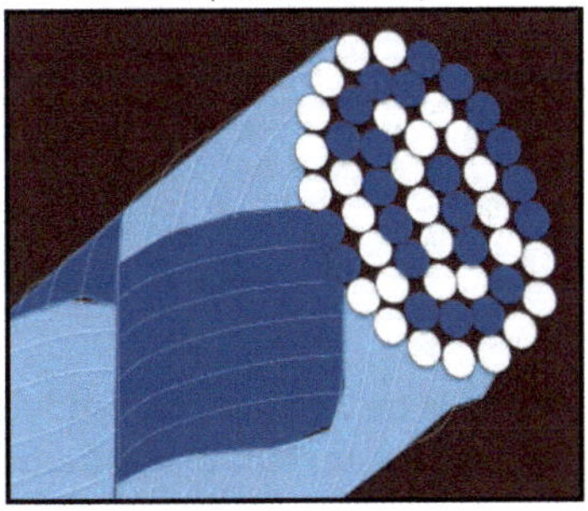

Capillarity (or wicking action)
Higher risk of infection
More drag
Lower tensile strength
Easier to grip
Secured knots hold well

Monofilament

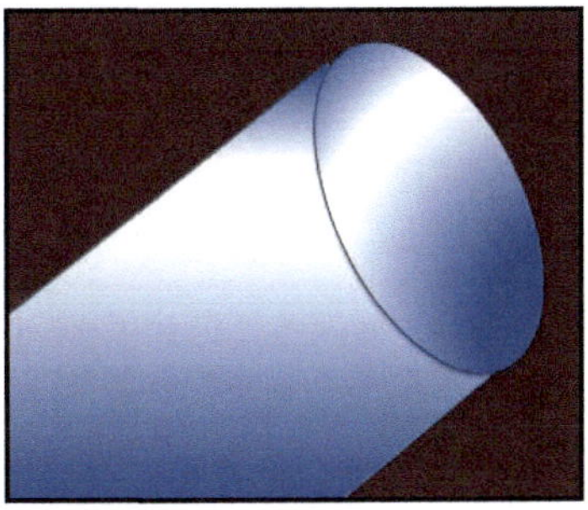

No capillarity
Lower risk of infection
Less drag
Greater tensile strength
Easier to grip
Greater memory effect of material*
Knots do not hold well

**The folded loops of thread, formed by packaging and storage (typically in a blister pack), tend to retain this shape during use and may cause accidental knotting or snagging when the working length of thread is not properly handled.*

Several features are used to describe a needle: its size, i.e., the length measured from the eye or the swaged (crimped insertion of thread) end of the needle to its point, the shape of the body, its curvature, and the form of the point.

The size is given in millimeters for conventional surgical sutures and in micrometers for microsurgical sutures. The majority of needles used today are swaged (crimped) onto a thread of the same diameter, the nature and length of which are specified on the package. Reusable and sterilizable needles can still be obtained from suppliers such as Unimed (Switzerland). The shape of its body and the form of its tip (point) define the needle. In cross section, the body is typically rounded, rectangular, or triangular. Points can be round, triangular, or trocar (tapercut, with three sharp cutting edges). The needle may combine identical or different body shapes and points, multiplying the assortment of needle types available.

Examples of Suture Needles Tips

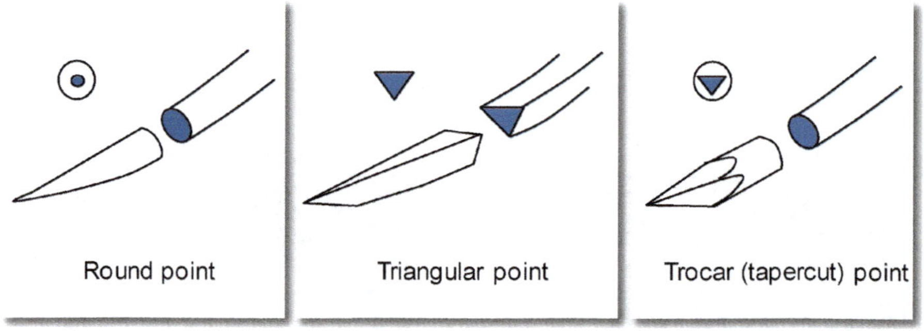

Suture selection is dependent on the anatomic site, surgeon's preference, and the required suture characteristics. The choice depends on the tissue, septic constraints, strength requirement, and durability in situ. Absorbable sutures can cause more inflammation than non-resorbable types. On the other hand, with animals that are difficult to handle or sensitive to stress, it is best to avoid having to remove the stitches after the skin has healed. In this case, a rapidly resorbable thread is preferable for skin suturing. The main consideration in needle selection is to minimize trauma. The length, diameter, and curvature of the needle influence the surgeon's ability to place a suture. A triangular point is preferable for thick skin and muscle tissue. The curvature of the needle is chosen according to the type of suture envisaged. For example, for a scalp suture, needle curvatures between 1/3, 3/8, and ½ are typically used.

Appendix 10: Procedure for Cresyl Violet Staining

Cresyl violet is a basic histological dye that stains purple the basophilic (acidic) cellular components DNA and RNA.

Once the tissue sections have been mounted on glass slides and thoroughly dried, prepare a solution of the dye by dissolving 5 g of cresyl violet acetate (Sigma, product n° 1791) in a mixture of 600 ml distilled water, 60 ml of 1 M sodium acetate (8.16 g sodium acetate in 60 ml of distilled water), and 340 ml of 1 M acetic acid (20.4 ml of concentrated acetic acid in 320 ml of distilled water). After stirring the solution for 48 h, filter. When the solution is ready, proceed with staining. This takes place in three stages:

- *Staining of Slides*
  - Soak the slides (rinse) in distilled water for 2 min.
  - Soak the slides in cresyl violet solution (preheated to 55 °C) for 9 min, checking the coloration of the tissue sections regularly since this amount of time may vary as a function of the staining power of the dye solution.
- *Dehydration of Slides: Soaking Steps*
  - 2 min in three successive baths of distilled water.
  - 2 min in 70 % ethanol.
  - 2 min in 95 % ethanol.
  - 30 s in an acid alcohol mixture (10 ml of concentrated acetic acid in 200 ml of 9 % ethanol).
  - 2 min in 95 % ethanol.
  - 2 min in two successive baths of 100 % ethanol.
  - 5 min in a solution of LMR© (mixture of purified paraffin and plastic polymers, used for protecting the tissue sections for several years against wrinkling and loss of transparency). LMR© replaces toluene and xylene that have typically been used at this step in the past.
- *Cover Slipping*

  When the stained slides are completely dry, mount the glass cover slips using Eukitt© (Sigma). Dry the slides for 2 h in a 37 °C oven or at room temperature for 24 h before microscopic observation.

GPSR Compliance

The European Union's (EU) General Product Safety Regulation (GPSR) is a set of rules that requires consumer products to be safe and our obligations to ensure this.

If you have any concerns about our products, you can contact us on ProductSafety@springernature.com

In case Publisher is established outside the EU, the EU authorized representative is:

Springer Nature Customer Service Center GmbH
Europaplatz 3
69115 Heidelberg, Germany

Batch number: 10755314

Printed by Printforce, the Netherlands